Agricultural Ecosystem Effects on Trace Gases and Global Climate Change

Agricultural Ecosystem Effects on Trace Gases and Global Climate Change

Proceedings of a symposium sponsored by Divisions A-3 and S-3 of the American Society of Agronomy and Soil Science Society of America. The symposium was held in Denver, CO, 28 October 1991.

Organizing Committee
Lowry A. Harper
Arvin R. Mosier
John M. Duxbury
Dennis E. Rolston, *chair*

Editorial Committee
Lowry A. Harper
Arvin R. Mosier
John M. Duxbury
Dennis E. Rolston, *chair*

Editor-in-Chief ASA
G. A. Peterson

Editor-in-Chief CSSA
P. S. Baenziger

Editor-in-Chief SSSA
R. J. Luxmoore

Managing Editor
David M. Kral

Associate-Editor-at-Large
Jon M. Bartels

ASA Special Publication Number 55

American Society of Agronomy, Inc.
Crop Science Society of America, Inc.
Soil Science Society of America, Inc.
Madison, Wisconsin, USA

1993

Cover Design: Chuck Andre

American Society of Agronomy, Inc.
Crop Science Society of America, Inc.
Soil Science Society of America, Inc.
677 South Segoe Road, Madison, WI 53711 USA

Library of Congress Cataloging-in-Publication Data

Agricultural ecosystem effects on trace gases and global climate change:
proceedings of a symposium / sponsored by Divisions A-3 and S-3
of the American Society of Agronomy and Soil Science Society of
America; organizing committee, Lowry A. Harper, Arvin R. Mosier,
John M. Duxbury; editorial committee, Dennis E. Rolston, Chair
... [et al.].
    p.   cm. — (ASA special publication; no. 55)
"The symposium was held in Denver, CO, 28 October 1991."
Includes bibliographical references.
ISBN 0-89118-113-X
    1. Agricultural ecology—Congresses. 2. Greenhouse gases—
Congresses. 3. Climatic changes—Congresses. I. Rolston, Dennis
Eugene, 1943–   . II. American Society of Agronomy. Division
A-3. III. Soil Science Society of America. Division S-3. IV. Series.
S1.A453 no. 55
[S589.7]
630 s—dc20
[628.5 '3]   92-45593
                                        CIP

Printed in the United States of America

CONTENTS

Foreword vii
Preface ix
Contributors xi
Conversion Factors for SI and non-SI Units xiii

1 Contributions of Agroecosystems to Global Climate Change
    John M. Duxbury, Lowry A. Harper,
    and A. R. Mosier 1

2 Methods for Measuring Atmospheric Gas Transport in
    Agricultural and Forest Systems
    O. T. Denmead and M. R. Raupach 19

3 Measurements of Greenhouse Gas Fluxes Using Aircraft- and
    Tower-based Techniques
    Raymond Desjardins, Philippe Rochette, Ian MacPherson,
    and Elizabeth Pattey 45

4 Use of Chamber Systems to Measure Trace Gas Fluxes
    G. L. Hutchinson and G. P. Livingston 63

5 Processes for Production and Consumption of Gaseous
    Nitrogen Oxides in Soil
    G. L. Hutchinson and Eric A. Davidson 79

6 Fluxes of Nitrous Oxide and Other Nitrogen Trace Gases from
    Intensively Managed Landscapes: A Global Perspective
    G. Philip Robertson 95

7 Cattle Grazing and Oak Trees as Factors Affecting Soil
    Emissions of Nitric Oxide from an Annual Grassland
    Eric A. Davidson, Donald J. Herman, Ayelet Schuster,
    and Mary K. Firestone 109

8 Denitrification in Subsurface Environments: Potential
    Source for Atmospheric Nitrous Oxide
    Charles W. Rice and Kirby L. Rogers 121

9 Nitrous Oxide Emissions and Methane Consumption in Wheat
    and Corn-Cropped Systems in Northeastern Colorado
    K. F. Bronson and A. R. Mosier 133

10 Methane: Processes of Production and Consumption
    Roger Knowles 145

11   Factors Affecting Methane Production in Flooded Rice Soils
      Charles W. Lindau, William H. Patrick, Jr.,
      and Ron D. DeLaune 157

12   Controls on Methane Flux from Terrestrial Ecosystems
      Joshua P. Schimel, Elisabeth A. Holland, and
      David Valentine 167

13   Methane Emissions from Flooded Rice Amended with a
      Green Manure
      Julie G. Lauren and John M. Duxbury 183

14   Rapid, Isotopic Analysis of Selected Soil Gases at
      Atmospheric Concentrations
      Paul D. Brooks, Gary J. Atkins, Donald J. Herman,
      Simon J. Prosser, and Andrew Barrie 193

Index 203

FOREWORD

The atmospheric concentration of the radiatively active trace gases, CO_2, CH_4, and N_2O, are increasing at the rate of about 0.5, 0.8, 0.3% per year, respectively. Biospheric interactions account for most of the removal of CO_2 from the atmosphere and even small changes in the inventory of C in living and dead materials above and below the soil surface can dramatically alter the atmospheric CO_2 budget. Methane and N_2O are primarily produced in the soil where soil management practices can exert control over the production and consumption mechanisms of these gases. Therefore it is critical that management procedures can be developed to minimize soils emissions of these gases.

Agronomy is now challenged with helping the earth to cope with a seemingly inevitable but highly unpredictable climate change resulting from release of radiatively active trace gases into the atmosphere. Technology, in the hands of agronomists and other scientists, can help us understand how agriculture can help us cope with or even ameliorate the progress of global change. In this publication, 34 leading scientists pool their knowledge to provide a reference relating the current state of knowledge concerning agroecosystems and radiatively active trace gases.

The American Society of Agronomy with its concern regarding agriculture and global climate change established a Workgroup on Global Climate Change to promote communication among the Tri-Societies of ASA and with others concerned with climate change. This book is sponsored by the ASA Workgroup along with symposia hosted at ASA meetings and other meetings.

D.N. DUVICK, *president*
American Society of Agronomy

G.H. HEICHEL, *president*
Crop Science Society of America

W.W. McFEE, *president*
Soil Science Society of America

PREFACE

Global climate change is an issue that has been thrust to the forefront of scientific, political, and general community interest. In the span of this human generation, the earth's climate is expected to change more rapidly than it has over any comparable period of recorded history. Some of the changes will result from natural processes, beyond human control, but much of this change is subject to anthropogenic influence arising from processes that are only beginning to be understood. Increasing concentrations of atmospheric radiatively active trace gases are being inadvertently affected by fossil fuel combustion; but other activities of industry, agriculture, forestry, changing land-use practices, waste disposal, and transportation also affect the chemical composition of the atmosphere. The measured and projected changes of the atmospheric concentrations of radiatively active trace gases have been modeled and estimated to predict changes in the global climate. Accuracy and reliability of these predictions are the subject of considerable debate among scientists and other concerned individuals, groups, and governmental agencies throughout the world. The objective of this book is to provide a review of current knowledge on the measurement of radiatively active trace gases in agricultural ecosystems and the effect of agriculture on the atmospheric concentrations of these gases.

This book is compiled from written papers presented at a symposium entitled, "Agroecosystem Effects on Radiatively Important Trace Gases and Global Climate Change," held at the American Society of Agronomy Meetings in Denver, CO, 27 Oct.–1 Nov. 1991. This symposium was cosponsored by Divisions A-3 and S-3. The book contains 14 chapters that represent the written contributions of the invited participants of the symposium and invited other contributions. Authors selected are recognized authorities in the field, both from within the ASA and SSSA Divisions as well as from outside the Societies. We hope that this book will prove useful to professionals and students in soil, plant, and atmospheric sciences and related environmental disciplines. Chapter 1 is an introduction to the effects of agricultural ecosystems on global climate change. Chapters 2 through 4 present discussions on experimental methods used to study fluxes of radiatively active trace gases, and chapters 6 through 13 discuss processes influencing the production and/or consumption of these gases. The last chapter covers the use of isotopes in measuring processes leading to the production and evolution of trace gases.

We appreciate the support of the ASA and SSSA that enabled the participation of authors from outside the USA in the publication of this book. The editors appreciate the careful and thoughtful efforts of the anonymous reviewers.

DENNIS E. ROLSTON, *chair*
Department of Land, Air, and Water Resources
University of California, Davis, CA

LOWRY A. HARPER, *editor*
Southern Piedmont Conservation Research Station
USDA-ARS, Watkinsville, GA

ARVIN R. MOSIER, *editor*
P.O. Box E
USDA-ARS, Fort Collins, CO

JOHN M. DUXBURY, *editor*
Department of Soils, Crops and Atmospheric Sciences
Cornell University, Ithaca, NY

CONTRIBUTORS

Gary J. Atkins

Development Scientist, Europa Scientific Ltd., Europa House, Electra Way, Crewe, Cheshire, CW1 1ZA, U.K.

Andrew Barrie

Director, Europa Scientific Ltd., Europa House, Electra Way, Crewe Cheshire, CW1 1ZA, U.K.

K. F. Bronson

Associate Visiting Scientist, Division of Soil and Water Science, International Rice Research Institute, Manila, Philippines

Paul D. Brooks

Spectroscopist, Department of Soil Science, University of California, Berkeley, CA 94720

Eric A. Davidson

Assistant Scientist, Woods Hole Research Center, Woods Hole, MA 02543

R. D. DeLaune

Professor, Wetland Biogeochemistry Institute, Louisiana State University, Baton Rouge, LA 70803-7511

O. T. Denmead

Chief Research Scientist, CSIRO Centre for Environmental Mechanics, Canberra ACT 2601, Australia

Raymond Desjardins

Micrometeorologist, Centre for Land and Biological Resources Research, Agriculture Canada, Central Experimental Farm, Ottawa, ON K1A 0C6, Canada

John M. Duxbury

Professor, Department of Soils, Crops and Atmospheric Sciences, Cornell University, Ithaca, NY 14853

Mary K. Firestone

Professor of Soil Science, Department of Soil Science, University of California, Berkeley, CA 94720

Lowry A. Harper

Soil Scientist, USDA-ARS, Southern Piedmont Conservation Research Center, Watkinsville, GA 30677

Donald J. Herman

Staff Research Associate, Department of Soil Science, University of California, Berkeley, CA 94720

Elisabeth A. Holland

Research Scientist, National Center for Atmospheric Research, Boulder, CO 80307

G. L. Hutchinson

Research Soil Scientist, USDA-ARS, Federal Building, Fort Collins, CO 80522

Roger Knowles

Professor of Microbiology, Department of Microbiology, Macdonald Campus, McGill University, Ste. Anne de Bellevue, PQ H9X 3V9, Canada

Julie G. Lauren

Postdoctoral Associate, Department of Soils, Crops and Atmospheric Sciences, Cornell University, Ithaca, NY 14853

Charles W. Lindau

Associate Professor, Wetland Biogeochemistry Institute, Louisiana State University, Baton Rouge, LA 70803

G. P. Livingston

Senior Scientist, TGS Technology, Inc., NASA Ames Research Center, Moffett Field, CA 94035-1000

Ian MacPherson — Senior Research Officer, National Research Council of Canada, Flight Research Laboratory, Ottawa, ON K1A 0R6, Canada

A.R. Mosier — Research Chemist, USDA-ARS, Federal Building, Fort Collins, CO 80522

William H. Patrick, Jr. — Boyd Professor and Director, Wetland Biogeochemistry Institute, Louisiana State University, Baton Rouge, LA 70803-7511

Elizabeth Pattey — Atmospheric Chemist, Agriculture Canada, Centre for Land and Biological Resources Research, Central Experimental Farm, Ottawa, ON K1A 0C6, Canada

Simon J. Prosser — Head of Development, Europa Scientific Ltd., Europa House, Electra Way, Crewe, Chesire, CW1 1ZA, U.K.

M.R. Raupach — Senior Principal Research Scientist, CSIRO Centre for Environmental Mechanics, Canberra ACT 2601, Australia

Charles W. Rice — Assistant Professor, Department of Agronomy, Kansas State University, Manhattan, KS 66506-5501

G. Philip Robertson — Associate Professor, W.K. Kellogg Biological Station and Department of Crop and Soil Sciences, Michigan State University, Hickory Corners, MI 49060

Philippe Rochette — Agrometeorologist, Agriculture Canada, Centre for Land and Biological Resources Research, Central Experimental Farm, Ottawa, ON K1A 0C6, Canada

Kirby L. Rogers — Graduate Research Assistant, Department of Agronomy, Kansas State University, Manhattan, KS 66506-5501

Dennis E. Rolston — Professor of Soil Science, Department of Land, Air, and Water Resources, University of California, Davis, CA 95616

Joshua P. Schimel — Assistant Professor of Microbial Ecology, Institute of Arctic Biology, University of Alaska, Fairbanks, AK 99775

Ayelet Schuster — Postdoctoral Research Associate, Department of Soil Science, University of California, Berkeley, CA 94720

David Valentine — Research Associate, Natural Resource Ecology Laboratory, Colorado State University, Fort Collins, CO 80523

Conversion Factors for SI and non-SI Units

Conversion Factors for SI and non-SI Units

To convert Column 1 into Column 2, multiply by	Column 1 SI Unit	Column 2 non-SI Unit	To convert Column 2 into Column 1, multiply by
Length			
0.621	kilometer, km (10^3 m)	mile, mi	1.609
1.094	meter, m	yard, yd	0.914
3.28	meter, m	foot, ft	0.304
1.0	micrometer, μm (10^{-6} m)	micron, μ	1.0
3.94×10^{-2}	millimeter, mm (10^{-3} m)	inch, in	25.4
10	nanometer, nm (10^{-9} m)	Angstrom, Å	0.1
Area			
2.47	hectare, ha	acre	0.405
247	square kilometer, km^2 (10^3 m)^2	acre	4.05×10^{-3}
0.386	square kilometer, km^2 (10^3 m)^2	square mile, mi^2	2.590
2.47×10^{-4}	square meter, m^2	acre	4.05×10^3
10.76	square meter, m^2	square foot, ft^2	9.29×10^{-2}
1.55×10^{-3}	square millimeter, mm^2 (10^{-3} m)^2	square inch, in^2	645
Volume			
9.73×10^{-3}	cubic meter, m^3	acre-inch	102.8
35.3	cubic meter, m^3	cubic foot, ft^3	2.83×10^{-2}
6.10×10^4	cubic meter, m^3	cubic inch, in^3	1.64×10^{-5}
2.84×10^{-2}	liter, L (10^{-3} m^3)	bushel, bu	35.24
1.057	liter, L (10^{-3} m^3)	quart (liquid), qt	0.946
3.53×10^{-2}	liter, L (10^{-3} m^3)	cubic foot, ft^3	28.3
0.265	liter, L (10^{-3} m^3)	gallon	3.78
33.78	liter, L (10^{-3} m^3)	ounce (fluid), oz	2.96×10^{-2}
2.11	liter, L (10^{-3} m^3)	pint (fluid), pt	0.473

Mass

To convert Column 1 into Column 2, multiply by	Column 1 SI Unit	Column 2 non-SI Unit	To convert Column 2 into Column 1, multiply by
2.20×10^{-3}	gram, g (10^{-3} kg)	pound, lb	454
3.52×10^{-2}	gram, g (10^{-3} kg)	ounce (avdp), oz	28.4
2.205	kilogram, kg	pound, lb	0.454
0.01	kilogram, kg	quintal (metric), q	100
1.10×10^{-3}	kilogram, kg	ton (2000 lb), ton	907
1.102	megagram, Mg (tonne)	ton (U.S.), ton	0.907
1.102	tonne, t	ton (U.S.), ton	0.907

Yield and Rate

To convert Column 1 into Column 2, multiply by	Column 1 SI Unit	Column 2 non-SI Unit	To convert Column 2 into Column 1, multiply by
0.893	kilogram per hectare, kg ha^{-1}	pound per acre, lb acre^{-1}	1.12
7.77×10^{-2}	kilogram per cubic meter, kg m^{-3}	pound per bushel, lb bu^{-1}	12.87
1.49×10^{-2}	kilogram per hectare, kg ha^{-1}	bushel per acre, 60 lb	67.19
1.59×10^{-2}	kilogram per hectare, kg ha^{-1}	bushel per acre, 56 lb	62.71
1.86×10^{-2}	kilogram per hectare, kg ha^{-1}	bushel per acre, 48 lb	53.75
0.107	liter per hectare, L ha^{-1}	gallon per acre	9.35
893	tonnes per hectare, t ha^{-1}	pound per acre, lb acre^{-1}	1.12×10^{-3}
893	megagram per hectare, Mg ha^{-1}	pound per acre, lb acre^{-1}	1.12×10^{-3}
0.446	megagram per hectare, Mg ha^{-1}	ton (2000 lb) per acre, ton acre^{-1}	2.24
2.24	meter per second, m s^{-1}	mile per hour	0.447

Specific Surface

To convert Column 1 into Column 2, multiply by	Column 1 SI Unit	Column 2 non-SI Unit	To convert Column 2 into Column 1, multiply by
10	square meter per kilogram, m^2 kg^{-1}	square centimeter per gram, cm^2 g^{-1}	0.1
1000	square meter per kilogram, m^2 kg^{-1}	square millimeter per gram, mm^2 g^{-1}	0.001

Pressure

To convert Column 1 into Column 2, multiply by	Column 1 SI Unit	Column 2 non-SI Unit	To convert Column 2 into Column 1, multiply by
9.90	megapascal, MPa (10^6 Pa)	atmosphere	0.101
10	megapascal, MPa (10^6 Pa)	bar	0.1
1.00	megagram per cubic meter, Mg m^{-3}	gram per cubic centimeter, g cm^{-3}	1.00
2.09×10^{-2}	pascal, Pa	pound per square foot, lb ft^{-2}	47.9
1.45×10^{-4}	pascal, Pa	pound per square inch, lb in^{-2}	6.90×10^3

(continued on next page)

Conversion Factors for SI and non-SI Units

To convert Column 1 into Column 2, multiply by	Column 1 SI Unit	Column 2 non-SI Unit	To convert Column 2 into Column 1, multiply by
Temperature			
$1.00\ (K - 273)$	Kelvin, K	Celsius, °C	$1.00\ (°C + 273)$
$(9/5\ °C) + 32$	Celsius, °C	Fahrenheit, °F	$5/9\ (°F - 32)$
Energy, Work, Quantity of Heat			
9.52×10^{-4}	joule, J	British thermal unit, Btu	1.05×10^{3}
0.239	joule, J	calorie, cal	4.19
10^{7}	joule, J	erg	10^{-7}
0.735	joule, J	foot-pound	1.36
2.387×10^{-5}	joule per square meter, J m^{-2}	calorie per square centimeter (langley)	4.19×10^{4}
10^{5}	newton, N	dyne	10^{-5}
1.43×10^{-3}	watt per square meter, W m^{-2}	calorie per square centimeter minute (irradiance), cal cm^{-2} min^{-1}	698
Transpiration and Photosynthesis			
3.60×10^{-2}	milligram per square meter second, mg m^{-2} s^{-1}	gram per square decimeter hour, g dm^{-2} h^{-1}	27.8
5.56×10^{-3}	milligram (H_2O) per square meter second, mg m^{-2} s^{-1}	micromole (H_2O) per square centimeter second, μmol cm^{-2} s^{-1}	180
10^{-4}	milligram per square meter second, mg m^{-2} s^{-1}	milligram per square centimeter second, mg cm^{-2} s^{-1}	10^{4}
35.97	milligram per square meter second, mg m^{-2} s^{-1}	milligram per square decimeter hour, mg dm^{-2} h^{-1}	2.78×10^{-2}
Plane Angle			
57.3	radian, rad	degrees (angle), °	1.75×10^{-2}

Electrical Conductivity, Electricity, and Magnetism

	Column 1 SI Unit	Column 2 non-SI Unit	
10	siemen per meter, S m^{-1}	millimho per centimeter, mmho cm^{-1}	0.1
10^4	tesla, T	gauss, G	10^{-4}

Water Measurement

	Column 1 SI Unit	Column 2 non-SI Unit	
9.73 × 10^{-3}	cubic meter, m^3	acre-inches, acre-in	102.8
9.81 × 10^{-3}	cubic meter per hour, m^3 h^{-1}	cubic feet per second, ft^3 s^{-1}	101.9
4.40	cubic meter per hour, m^3 h^{-1}	U.S. gallons per minute, gal min^{-1}	0.227
8.11	hectare-meters, ha-m	acre-feet, acre-ft	0.123
97.28	hectare-meters, ha-m	acre-inches, acre-in	1.03 × 10^{-2}
8.1 × 10^{-2}	hectare-centimeters, ha-cm	acre-feet, acre-ft	12.33

Concentrations

	Column 1 SI Unit	Column 2 non-SI Unit	
1	centimole per kilogram, cmol kg^{-1} (ion exchange capacity)	milliequivalents per 100 grams, meq 100 g^{-1}	1
0.1	gram per kilogram, g kg^{-1}	percent, %	10
1	milligram per kilogram, mg kg^{-1}	parts per million, ppm	1

Radioactivity

	Column 1 SI Unit	Column 2 non-SI Unit	
2.7 × 10^{-11}	becquerel, Bq	curie, Ci	3.7 × 10^{10}
2.7 × 10^{-2}	becquerel per kilogram, Bq kg^{-1}	picocurie per gram, pCi g^{-1}	37
100	gray, Gy (absorbed dose)	rad, rd	0.01
100	sievert, Sv (equivalent dose)	rem (roentgen equivalent man)	0.01

Plant Nutrient Conversion

	Elemental	Oxide	
2.29	P	P$_2$O$_5$	0.437
1.20	K	K$_2$O	0.830
1.39	Ca	CaO	0.715
1.66	Mg	MgO	0.602

1   Contributions of Agroecosystems to Global Climate Change

John M. Duxbury

Department of Soil, Crop and Atmospheric Sciences
Cornell University
Ithaca, New York

Lowry A. Harper

USDA-ARS
Watkinsville, Georgia

A. R. Mosier

USDA-ARS
Fort Collins, Colorado

Climate change has been a natural feature of the Earth's past. Average global surface temperatures have varied by 5 to 7 °C over glacial–interglacial cycles (100 000-yr time scale) and by 2 °C in the 10 000 yr since the last ice age (Folland et al., 1990). Changes in the Earth's climate due to anthropogenically induced increases in the atmospheric greenhouse effect are anticipated in the near future. Average global surface temperatures have, in fact, increased by 0.45 ± 0.15 °C in the last century but linkage of this change to anthropogenic activities is controversial (Folland et al., 1990). Simulation of climate through atmospheric general circulation models (GCM's) is the principal methodology by which anthropogenic effects on climate are evaluated and projected. Much attention has been given to prediction of the climate effects of doubling the atmospheric CO_2 level since this could occur by the middle of the next century. For this situation, the various GCM's predict an average global warming between 1.5 and 4.5 °C, and changes in global rainfall amounts and distribution (Bretherton et al., 1990; Mitchell et al., 1990). Furthermore, the combined effect of increases in other trace gases, CH_4, N_2O, chlorofluorocarbons, and ozone (O_3) presently contribute about as much to greenhouse forcing in the atmosphere as does CO_2.

The purpose of this chapter is to introduce the topic of greenhouse gases and climate change, to summarize and place in perspective present knowledge of agricultural contributions to the greenhouse effect, and to discuss how these may change in the future.

THE GREENHOUSE EFFECT

The Earth's climate is controlled by the radiative balance of the atmosphere (Fig. 1-1). The Earth absorbs about one-half of the shortwave radiation it intercepts and emits longer wave thermal infrared radiation from its surface. On a long-term basis the energy of incoming radiation is balanced by that of outgoing radiation. Perturbations to the energy balance alter climate, that is, they create radiative forcing of climate. There are several ways in which perturbations to the energy balance can occur, including variation in solar radiation, variation in the Earth's orbit, and changes in the greenhouse effect. The greenhouse effect occurs because some of the atmospheric trace gases, notably water vapor and CO_2, partially absorb infrared radiation coming from the Earth's surface, leading to a warming of the atmosphere. The heating of the atmosphere represents a small net absorption of energy by the Earth's climate system, but equilibrium between energy input and output is quickly re-established as a hotter atmosphere enhances energy transfer to space.

The importance of the greenhouse effect to average surface temperatures on different planets is shown in Table 1-1 (IPCC, 1990). On Earth, the combination of about 1% water vapor, 0.04% CO_2, and smaller amounts of other greenhouse gases in the atmosphere warms the surface by 33 °C. Water vapor accounts for about 60 to 70% of the warming and CO_2 for about 25% of the warming. Venus and Mars have atmospheres containing mostly CO_2 and atmospheric pressures much higher and lower than the Earth, respectively. Consequently, the greenhouse effect on Venus is about 520 °C, whereas that on Mars is only 10 °C.

The current concern over Earth's climate centers on the probable enhancement of the greenhouse effect due to anthropogenically related emissions of greenhouse gases. Industrial and agricultural operations have increased emissions of several naturally occurring greenhouse gases (CO_2, CH_4, and N_2O) with new greenhouse gases, the chlorofluorocarbons

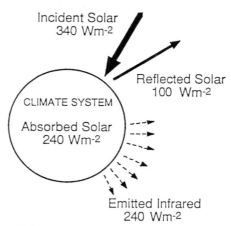

Fig. 1-1. The Earth's radiative energy balance.

Table 1-1. The greenhouse effect on different planets.

	Surface pressure relative to earth	Main greenhouse gases	Surface temperature without greenhouse effect	Observed surface temperature	Greenhouse warming
			°C		
Venus	90	>90% CO_2	−46	477	523
Earth	1	0.035% CO_2 1% H_2O	−18	15	33
Mars	0.007	>80% CO_2	−57	−47	10

(CFC's), also being added to the atmosphere. Additionally, fluxes of other gases to and from natural and agricultural ecosystems, including ammonia (NH_3), N oxides (NO_x), nonmethane hydrocarbons (NMHC's), S dioxide (SO_2) and various organosulfur compounds have been altered and can indirectly affect atmospheric concentrations of greenhouse gases through their impacts on atmospheric chemistry and/or ecosystem functioning (Mooney et al., 1987).

CHANGES IN GREENHOUSE GAS LEVELS

Atmospheric increases in CO_2, CH_4, N_2O and CCl_3F (CFC-11) since 1750 are shown in Fig. 1-2. Data prior to the 1950s were obtained from analysis of gas bubbles entrapped in Antarctic ice. Systematic and contemporary

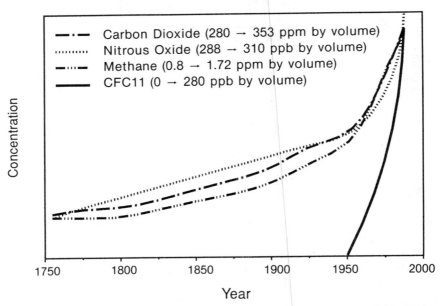

Fig. 1-2. Atmospheric increases in CO_2, CH_4, N_2O and CFC-11 since 1750 (IPCC, 1990).

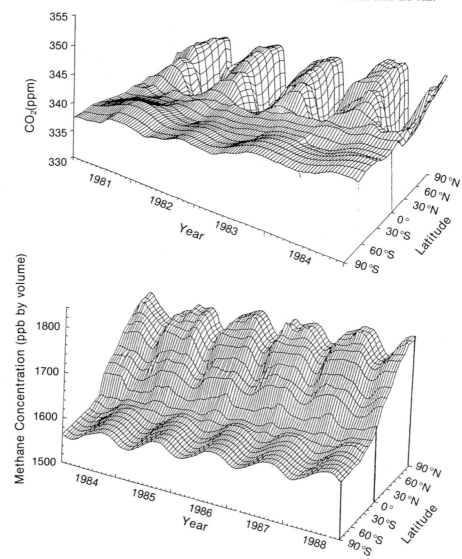

Fig. 1-3. Global atmospheric distribution of CO_2 and CH_4 (Conway et al., 1988; Watson et al., 1990).

measurements of atmospheric CO_2 levels began in 1958 with Keeling's work at Mauna Loa (Keeling, 1986). More comprehensive measurements of a suite of trace gases were initiated in 1977 with the National Oceanographic and Atmospheric Administration's (NOAA) geographic monitoring network for climate change (GMCC; Elkins & Rossen, 1989). Atmospheric concentrations of CO_2, CH_4, and N_2O increased slowly between 1750 and 1950, then increased rather abruptly between 1950 and present. The concentration of

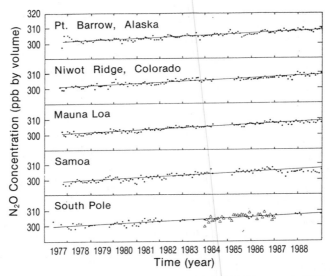

Fig. 1–4. Atmospheric levels of N₂O from the NOAA–GMCC network (Elkins & Rossen, 1989).

CFC-11, which is contemporary and exclusively anthropogenic, has increased dramatically since its introduction in about 1950.

The recent global distribution of atmospheric CO_2 (Conway et al., 1988) and CH_4 (Steele et al., 1987), reproduced in Fig. 1–3, shows: (i) a general increase in concentration with time, (ii) higher levels in the northern hemisphere associated with greater land mass and source strength in this hemisphere, and (iii) seasonal fluctuations that are caused by climate-controlled cycles of production and consumption. The principal driving force for the CO_2 pattern is photosynthetic uptake, whereas the CH_4 pattern is thought to be controlled by variations in both biological production and oxidation in the atmosphere by reaction with ·OH radicals (Watson et al., 1990). In contrast, neither hemispheric differences nor seasonal fluctuations are apparent from recent measurements of atmospheric N_2O (Elkins & Rossen, 1989; Fig. 1–4), suggesting different controls on its level.

GLOBAL WARMING POTENTIALS AND RADIATIVE FORCING OF CLIMATE

A single index, called the Global Warming Potential (GWP), has been developed to compare the relative greenhouse effects of different trace gases (Shine et al., 1990). The GWP concept combines the capacity of a gas to absorb infrared radiation, its residence time in the atmosphere, and a time frame over which climate effects are to be evaluated. Preliminary estimates of GWP's for various trace gases are shown relative to CO_2 for two differ-

Table 1–2. Global warming potential (GWP) of various trace gases.

Gas	Residence time	Relative absorption capacity†	Relative GWP‡	
			20 yr	500 yr
	yr	unit mass		
CO_2	120	1	1	1
CH_4				
Direct§	10	58	26	2
Indirect			37¶	7
N_2O	150	206	270	190
CCl_3F (CFC-11)	60	3970	4500	1500
CCl_2F_2 (CFC-12)	120	5750	7100	4500
$CHClF_2$ (HCFC-22)	15	5440	4100	510
$CHCl_2CF_3$ (HCFC-123)#	1.6	2860	310	29

† Per unit mass change from present concentrations.
‡ Following addition of 1 kg of each gas.
§ Direct = through direct radiative effect of CH_4; indirect = through the participation of CH_4 in chemical processes which lead to the formation of other radiatively active species.
¶ Estimated contributions of 24, 10, and 3 for generation of tropospheric O_3, stratospheric H_2O, and CO_2, respectively.
# One of the proposed CFC substitutes.

ent periods in Table 1–2, which is adapted from Shine et al. (1990). The calculations were made for a 1-kg addition of each trace gas to the ambient atmosphere. They are adjusted for any generation of secondary greenhouse gases formed from destruction of the primary gas but do not include effects of CFC's and N_2O through depletion of stratospheric O_3. Furthermore, no corrections are made for overlapping of absorption bands among the various trace gases; and several assumptions made, for example, that the atmospheric lifetimes of the gases will remain constant over time, are probably incorrect. Although the process of calculating GWP's is complex and imperfect at present, the results serve to illustrate important differences between the trace gases. Methane, which has a much shorter atmospheric residence time than CO_2 (10 vs. 120 yr), has a 60-fold higher GWP in a 20-yr time frame due partly to its greater absorptive capacity (direct effect) and partly to the generation of CO_2, tropospheric O_3, and stratospheric water vapor from its atmospheric chemistry (indirect effects). Over a 500-yr time frame, CH_4 still has a ninefold larger GWP than CO_2, which is mostly due to indirect effects rather than to CH_4 directly. Nitrous oxide has an atmospheric residence time similar to CO_2 but a 200-fold greater absorptive capacity and a correspondingly higher GWP than CO_2. It is also a more effective greenhouse gas than CH_4, especially on the longer-term basis. The CFC's and HCFC's have absorption capacities from one to over three orders of magnitude greater than the other trace gases, but generally similar residence times so all are extremely potent greenhouse gases. Of several possible CFC substitutes, $CHCl_2CF_3$ (HCFC-123) has the lowest GWP, primarily because of its relatively short atmospheric residence time of 1.6 yr.

Contributions of the different trace gases to radiative forcing of climate in the period from 1765 to 1990 are shown in Fig. 1–5 (Shine et al., 1990).

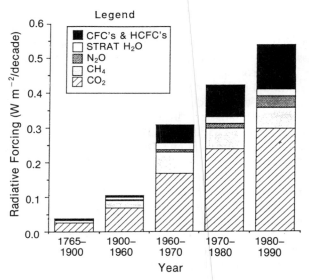

Fig. 1-5. Historical contributions of greenhouse gases to radiative forcing of climate (Shine et al., 1990).

Increases in atmospheric CO_2 levels have been, and continue to be, the main contributor to radiative forcing, accounting for 61% of the overall change and 56% in the last decade. Since 1960, the CFC's and N_2O have made progressively larger contributions, whereas the effect of CH_4 has remained more or less constant.

AGRICULTURAL CONTRIBUTIONS TO TRACE GAS EMISSIONS AND GLOBAL WARMING

The present contributions of different anthropogenic activities to global warming are shown in Fig. 1-6 (USEPA, 1989). While fossil fuel use and

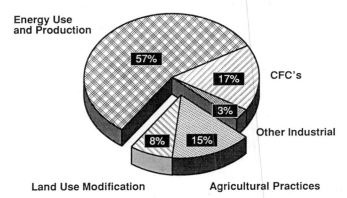

Fig. 1-6. Present contributions of anthropogenic activities to global warming (USEPA, 1990).

activities associated with its extraction and transport are major contributors, agricultural activities account for about one-quarter of the total effect. Two-thirds of the agricultural contribution is related to agricultural practices and one-third is associated with conversion of land to agricultural use, principally deforestation in the tropics.

Estimated annual budgets for CO_2, N_2O, and CH_4 and the contributions of agriculture to these budgets are presented in Table 1-3. Anthropogenic activities are much more important sources of CH_4 and N_2O than they are of CO_2, accounting for 64, 24, and 3%, respectively, of total emissions of these gases. Although the net annual flux, determined by subtracting total sink strength from total source strength, does not balance the observed rate of increase in atmospheric content for any of the gases, it must be recognized that there is considerable uncertainty in many of the source–sink estimates. For CO_2, the anthropogenically derived component is only 3% of total sources but emissions of CO_2 from fossil fuel combustion, the major anthropogenic source, are well documented. Therefore, it is clear that either the agricultural contribution associated with deforestation and loss of soil organic matter has been overestimated or that a sink has been underestimated or not identified.

Houghton and Skole (1990) and Houghton et al. (1983) used a book-keeping approach to estimate that cumulative release of CO_2 due to conversion of land to agriculture between 1850 and 1985 was 117 ± 35 Gt C. This compares with the cumulative release of 195 ± 20 Gt C from fossil fuel consumption in this same period (Watson et al., 1990). Houghton et al. (1983) attributed 80% (94 Gt C) of the decreased C storage to loss of trees and 20% (23 Gt C) to loss of soil organic matter. Although anthropogenic activities related to agriculture have contributed significantly to rising atmospheric CO_2 levels in the last 150 yr, the major anthropogenic flux today comes from fossil fuel combustion, which added 5.7 Gt CO_2–C to the atmosphere in 1987. Current release of CO_2 due to conversion of land to agriculture is estimated to be between 0.6 to 2.6 Gt C yr^{-1} (Detwiler & Hall, 1988; Houghton et al., 1987) with a value of 1.6 Gt C yr^{-1} often being used (Watson et al., 1990).

In contrast, several dynamic simulation models suggest that current losses of C due to land conversion to agriculture are balanced by increased net primary productivity (NPP) in native terrestrial ecosystems, which leads to greater C storage in both aboveground biomass and soil organic matter. Increased NPP is postulated to be due to higher atmospheric CO_2 levels, i.e., to a CO_2 fertilization effect (Goudriaan & Ketner, 1984; Esser, 1987, 1990), although it would seem that increased inputs of nutrients, particularly N could also be a factor. Proposed fertilization effects are, however, difficult to evaluate because the terrestrial biomass and soil organic matter pool sizes are 250 to 750 times larger than the imbalance in the annual C budget. Nevertheless, recent analyses by Tans et al. (1989, 1990) suggest a northern temperate region terrestrial sink for CO_2. The oceans, which are the major buffer of atmospheric CO_2 and which would eventually absorb almost all of the additional CO_2 introduced into the atmosphere, only remove about 30% of that

Table 1-3. Estimated annual budgets for CO_2, CH_4, and N_2O.†

	CO_2 (10^{15} g C yr^{-1})		CH_4 (10^{12} g C yr^{-1})		N_2O (10^{12} g N yr^{-1})	
Sources						
Natural	Plant respiration	60	Wetlands	85	Ocean release	2.0
	Litter/soil organic matter decomposition	60	Oceans/freshwater	10	Tropical forests	3.0
	Ocean release	105	Wild animals/termites	30	Other forests	1.1
			Biomass burning	20	Grassland	1.6
					Biomass burning	0.1
	Sub-total	225		145		7.8
Anthropogenic	Fossil fuel combustion	5.0	Rice paddies	85	Fertilizer use	0.7
	Vegetation/soil organic matter destruction	1.8	Domestic animals	60	Grassland	0.7
			Landfills	30	Increase in cultivated land	0.7
			Mining/natural gas	60	Biomass burning	0.1
			Biomass burning	20	Combustion	0.2
	Sub-total	6.8		255		2.4
Total sources		231.8		400		10.2
Anthropogenic (% total)		3		64		24
Agricultural (% anthropogenic)		26		65		92
Sinks						
	Ocean uptake	106.6	Reaction with OH	375	Photolysis in stratosphere	10.5
	Ocean burial	0.1	Oxidation by soil	26		
	Plant photosynthesis	120				
Total sinks		226.7		401		10.5
Net change (sources−sinks)		5.1		−1		−0.3
Atmospheric increase		3.0		40		3.7
Imbalance		2.1 (sink)		41 (source)		4.0 (source)

† Constructed from data in Bouwman (1990), Schlesinger (1990), USEPA (1990), Watson et al. (1990), Ojima et al. (1992).

generated annually from anthropogenic sources because of slow exchange between surface waters and the deep ocean (Tans et al., 1990).

Anthropogenic sources account for 64% of total CH_4 emissions. Agriculture and related activities contribute two-thirds of the anthropogenic source or 41% of all sources. This production is offset by oxidation of CH_4, which largely takes place in the atmosphere through reaction with the $\cdot OH$ radical (Cicerone & Oremland, 1988). Since CH_4 source budgets are often "balanced" against destruction, the recent discovery that the rate coefficient for the reaction of $\cdot OH$ with CH_4 is about one-fourth less than previously thought brings considerable uncertainty to the global atmospheric CH_4 budget (Vaghjiani & Ravishankara, 1991). Additionally, the recent observation that CH_4 is consumed in soils introduces a new term for consideration in the global CH_4 budget (Mosier & Schimel, 1991; Mosier et al., 1991; Steudler et al., 1989; Whalen & Reeburgh, 1990). Ojima et al. (1992) estimate a current global terrestrial CH_4 sink of 20 to 33 Tg CH_4–C yr^{-1} or between 5 to 8% of total sources. Because the annual production of tropospheric $\cdot OH$ is almost all consumed by reaction with CH_4 and CO (Watson et al., 1990), it is possible that further increases in CH_4 emissions will saturate the atmospheric capacity for its oxidation, leading to greater persistence of CH_4. Ojima et al. (1992) suggest that increases in the concentration of atmospheric CH_4 will increase the importance of the soil CH_4 sink.

The annual budget for N_2O appears to underestimate sources, although high spatial and temporal variability make these particularly difficult to estimate (Duxbury et al., 1982; Eichner, 1990). Tropical forests are estimated to contribute 40% of natural sources, but the data base is somewhat limited (Keller et al., 1986; Livingston et al., 1988). More rapid N cycling in the tropical environment is considered to cause higher N_2O emissions than found in cooler climates (Matson & Vitousek, 1987). Anthropogenic activities appear to have increased N_2O emissions by about one-third of the baseline value, and increased emissions are almost totally associated with agriculture. Budgets for N_2O are, however, often fragmentary; for example, the IPCC budget (Watson et al., 1990) includes values for forested soils and fertilizer but neglects to include any value for grasslands, although the global area of this biome is almost as great as that of forestland (3.1×10^9 ha vs. 4.0×10^9 ha). No contribution from increased biological N fixation (BNF) through agricultural systems is included, nor is the impact of agricultural use of land per se considered. By analogy to the tropical environment, increased N cycling in agricultural systems could lead to higher N_2O emissions compared to the natural ecosystems they replaced.

It is also likely that N_2O production resulting from fertilizer and increased use of BNF is underestimated because the effect of a N input is usually only partially traced through the environment. Figure 1–7 illustrates some of the flows of N following input of 100 kg of fertilizer. What can be considered the primary cycle is shown by the dashed lines. In this example, 50 of the 100 kg are harvested in the crop and 50 are lost by the combination of leaching (25), surface run-off (5), and volatilization (20, primarily denitrifi-

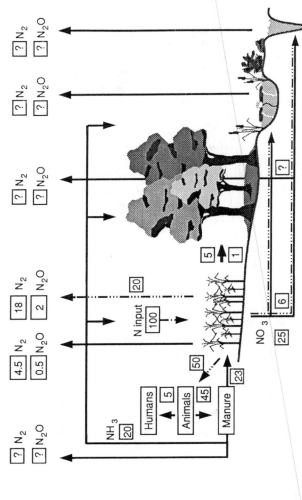

Fig. 1-7. A simplified flow of fertilizer N through the environment.

cation). If N_2O comprises 10% of the volatilized N, 2 kg N_2O-N would be generated in the primary cycle. If assessment of the fertilizer effect stops at this point, and most do, only 20 of the 100 kg N has been returned to the atmosphere, yet a reasonable assumption is that almost all of it would be returned within a time frame of a few years. Secondary flows shown by the solid lines include feeding of the 50 kg of harvested N to animals, which generate 45 kg of manure N. The manure is returned to cropland to create a secondary crop cycle, however, about one-half of the N in manure is volatilized as NH_3 prior to or during manure application. Volatilized NH_3 is aerially dispersed and subsequently returned to and cycled through both natural ecosystems and cropland. Ammonia volatilization from agricultural systems is globally important but its impact on N_2O emissions has not been explicitly addressed. In U.S. agriculture, the annual production of animal manure N is equal to the annual use of fertilizer N (Bouldin et al., 1984) and probably 30 of the 80 Tg fertilizer N per year used globally are volatilized as NH_3.

Similarly, the amount of N_2O arising from leached NO_3^-, which may average 20 to 25% of applied N, is not known but much may be denitrified in riparian zones or cycled through wetland or aquatic vegetation. A complete accounting of fertilizer and biologically fixed N is difficult to achieve but needed, if we are to accurately assess the impact of increased use of N in agricultural ecosystems on terrestrial N_2O emissions.

DIFFERENTIAL EFFECTS OF AGRICULTURE ON TRACE GAS FLUXES

Agricultural activities or practices can lead to opposing effects on fluxes of different greenhouse gases from both agricultural and natural ecosystems. Several effects are associated with N fertilization. Nitrogen additions to forest soils led to a combined two- to threefold increase in emissions of COS and CS_2 (Melillo & Steudler, 1989), which could create a cooling effect (see following discussion on aerosols). At the same time, oxidation of CH_4 by soil microorganisms was reduced by 30% (Steudler et al., 1989), which could create a warming effect. Oxidation of CH_4 in native and cultivated grassland soils is similarly affected by N fertilization (Mosier et al., 1991). To our knowledge, potential effects of N fertilization on carbonyl sulfide (COS) and carbon disulfide (CS_2) emissions from agricultural soils have not been investigated.

In rice (*Oryza sativa*) paddies, a tradeoff between CH_4 and N_2O emissions may well occur depending on water management. Conditions that favor CH_4 generation promote reduction of N_2O to N_2, whereas less reduced conditions, which may reduce CH_4 generation, would likely allow more N_2O to escape to the atmosphere. Periodic draining, for example, would likely increase nitrification and NO_3 availability for denitrification upon reflooding. Such interactions need to be addressed when developing water management strategies for mitigation of CH_4 emissions from paddy rice.

OTHER TRACE GASES, AEROSOLS, AND CLOUDS

While the contributions of industrial and agricultural activities to rising atmospheric levels of CO_2, CH_4, and N_2O and hence to the greenhouse effect are relatively well documented, the quantitative impacts of other trace gases and atmospheric constituents are less certain. The extent and impact of changing O_3 levels in both the stratosphere and the troposphere are the most difficult to evaluate. Reductions in stratospheric O_3, which now routinely occur over the Antarctic every spring and may be more widespread, change processes that have opposing effects on global climate. Reduced stratospheric O_3 allows greater penetration of solar radiation into the troposphere and hence surface warming; however, this effect is countered by reduced emission of solar and long-wave radiation from the stratosphere to the troposphere. These effects are considered to be similar in magnitude but they are dynamic, heterogeneous over the globe, and incompletely understood (Watson et al., 1990).

Tropospheric O_3 levels are controlled by a complex set of reactions and feedbacks that are also temporally and spatially variable. Ozone is generated in the troposphere by a series of reactions involving photo-oxidation of CO, CH_4, and NMHC's in the presence of NO_x, a common anthropogenic pollutant. Reactant concentrations and reaction outcomes however, are variable; at NO_x concentrations $< \sim 10$ ppt$_v$, the net result is actually a decrease in tropospheric O_3 and $\cdot OH$, whereas production of O_3 and $\cdot OH$ occurs at higher NO_x levels (Crutzen, 1988). Additionally, feedbacks between O_3, its precursor gases, and $\cdot OH$ introduces self-regulation of both O_3 and CH_4. Elevated levels of O_3 and NO_x enhance $\cdot OH$ production, which in turn oxidizes CO, CH_4, and NMHC's with consequent reduction in O_3 generation. Current evidence suggests that levels of both tropospheric O_3 and CO are increasing at about 1% yr^{-1} (Watson et al., 1990). The lack of systematic global data on atmospheric concentrations of NO_x and NMHC's prevents evaluation of temporal trends, however, the large known combustion sources of NO_x (Galbally, 1989) and regionally elevated atmospheric concentrations (Watson et al., 1990) suggest that its atmospheric concentration should be increasing globally.

Evaluation of both the direct and indirect contributions of O_3 to the greenhouse effect cannot be made with confidence because of insufficient knowledge about its distribution and behavior in the atmosphere. Ozone does, however, have a radiative absorption potential $2000\times$ higher than that of CO_2. A preliminary assessment suggests that tropospheric O_3 may have contributed about 10% of the total radiative forcing since preindustrial times (Shine et al., 1990), but this may have been counteracted by cooling due to decreases in O_3 in the lower stratosphere (Hansen et al., 1989; Lacis et al., 1990).

Aerosols, which are solids or liquid droplets having diameters in the 0.002- to 20-μm range, influence the Earth's climate system directly through absorption and scattering of both solar and thermal infrared radiation and indirectly through modification of the physical properties of clouds (Watson et al., 1990; Shine et al., 1990). Tropospheric aerosols have lifetimes of days to weeks and presumably are concentrated near the sources. They include dust, trace metals, H_2SO_4 and SO_4, HNO_3 and NO_3, organic materials, and elemental C. Stratosphere aerosols are largely created by volcanic eruptions, which inject dust and SO_2 into the stratosphere, and by oxidation of (COS) to yield sulfate. They have lifetimes to a few years. Analysis of Greenland ice cores show that tropospheric aerosol levels have been increasing since the industrial revolution (Neftel et al., 1985; Mayewsky et al., 1986) but the atmospheric record is insufficient to directly document changes beyond industrialized regions (Charlson, 1988). Concentrations of stratospheric aerosols can be substantially altered for several years following volcanic eruptions, and effects due to any changes in COS flux would be difficult to discern. Carbonyl sulfide and CS_2, a precursor gas, are generated biologically in soils, especially under anaerobic conditions (Minami & Fukushi, 1981a,b; Steudler & Peterson, 1984; Khalil & Rasmussen, 1984; Servant, 1989). The effects of agriculture, especially the possible contribution of rice paddies, on the fluxes of these gases has not been investigated.

The climate effects of aerosols are also difficult to assess and can be either to heat or cool depending upon aerosol physical and chemical characteristics and a number of local conditions (Shine et al., 1990). Tropospheric aerosols act as condensation nuclei for the formation of clouds from water vapor and increase the reflectivity of clouds, which leads to a cooling effect. Again, however, gaps in knowledge and lack of data prevent overall conclusions from being made at present.

PROJECTED CHANGES IN TRACE GAS EMISSIONS ASSOCIATED WITH AGRICULTURE AND OPPORTUNITIES FOR MITIGATION

Deforestation in developing, tropical countries will most likely continue as their populations grow and will lead to continuing land conversion releases of CO_2 to the atmosphere. Much attention is currently being given to the development of low input sustainable agricultural systems for resource-poor farmers in developing countries, however, this approach promotes extensification of agriculture. From a global climate perspective, it would be better to intensify agriculture so that more forestlands remain. At the same time, adoption of practices to promote C storage, or to prevent its loss, in terrestrial ecosystems should be encouraged. Replanting of forests has been suggested as a way to counteract fossil fuel release of CO_2, but the scale required for this to be effective is daunting. For example, Wiersum and Ketner (1989) estimate that an area of 200×10^6 ha of new forests could sequester 0.6 Gt C yr^{-1} over a 30-yr productive life and that 30×10^6 ha of fertilized woodland could sequester 0.14 Gt C yr^{-1} during 8-yr rotations thereby substituting for 0.12 Gt yr^{-1} of fossil fuel C. The combined 230×10^6 ha would cover one-half the area of Europe and annually would sequester only 15% of the CO_2-C presently generated by combustion of fossil fuel.

Storage of C in soil organic matter also has been considered, and the only practical way to positively influence this is to switch to conservation or no-till practices. An analysis of the impact of increasing no-tillage cropland in the USA from the present 27 to 76% of agricultural land shows a net gain of 0.42 Gt C in soil organic matter and a savings of 0.01 Gt C in fossil fuel use over the next 30 yr (Kern & Johnson, 1991). The combined savings of 0.43 Gt C is equivalent to 1.1% of projected total fossil fuel use in the USA in the same time period. The conclusion from these analyses is clear—that although some sequestration and savings of C can be achieved through forest planting and changes in agricultural practices, these are not the best scientific ways to approach the problem of CO_2 release from fossil fuel combustion.

Global population growth in the next 30 yr is expected to increase the demand for rice from 480 to 780 $\times$ 10^6 t, a 65% increase. It is anticipated that this demand will be met by increasing the area of paddy rice and by increasing yield through use of organic inputs, improved fertilizer efficiency and development of higher yielding cultivars (Braatz & Hogan, 1991; USEPA, 1990). Increased acreage of paddy and use of organic inputs, including legume green manures to supply N, will lead to increased CH_4 emissions. Better crop growth may also lead to higher CH_4 emissions if more C is cycled into soil through the root system. If the projected increase in rice production is accompanied by a proportional increase in CH_4 emissions and there is no change in other trace gas concentrations, the contribution of rice paddies to global radiative forcing would increase from 3 to 5%.

On the other hand, a 10% reduction in current CH_4 emissions from rice would stabilize the atmospheric load from this source. Possible opportunities for mitigation are essentially still at the research stage, however, approaches such as cultivar selection, N source and placement options, use of nitrification inhibitors, and alternative water management strategies appear to hold some promise.

Nitrous oxide emissions from agricultural lands are also likely to increase in the future because of increases in both the extent of cultivated land and fertilizer N inputs. Crop land area in the developing countries (excluding China) is expected to increase from 600 $\times$ 10^6 to 950 $\times$ 10^6 ha by 2025 (USEPA, 1990), but little change is anticipated for developed countries. Fertilizer N inputs are projected to increase from 80 to 140 Tg N yr^{-1} in this same time frame, with almost all of the increase occurring in developing countries. Without mitigation, it is likely that these conditions could lead to doubling of the agricultural contribution to global N_2O flux. Potential mitigation strategies focus on improved use of N in agriculture; available technologies include avoidance of overfertilization, more timely application of N, provision for continuous plant cover so that surplus fertilizer N and that mineralized from soil organic matter when crops are not normally present is recycled rather than leached, and use of N sources that minimize N_2O emissions. Research may also lead to improvements in fertilizer N efficiency and manure handling procedures that reduce NH_3 volatilization.

REFERENCES

Bouldin, D.R., S.D. Klausner, and W.S. Reid. 1984. Use of nitrogen from manure. p. 221–245. *In* R.D. Hauck (ed.) Nitrogen in crop production. ASA, CSSA, and SSSA, Madison, WI.

Bouwman, A.F. 1990. Exchange of greenhouse gases between terrestrial ecosystems and the atmosphere. p. 61–127. *In* A.F. Bowman (ed.) Soils and the greenhouse effect. John Wiley, Chichester.

Braatz, B.V., and K.B. Hogan. 1991. Sustainable rice productivity and methane reduction research plan. USEPA, Washington, DC. (In press.)

Bretherton, F.P., K. Bryan, and J.D. Woods. 1990. Time-dependent greenhouse gas induced climate change. p. 179–193. *In* J.T. Houghton et al. (ed.) Climate change: The IPCC scientific assessment. Cambridge Univ. Press, Cambridge.

Charlson, R.J. 1988. Have concentrations of tropospheric aerosols changed? *In* F.S. Rowland and I.S.A. Isaksen (ed.) The changing atmosphere. John Wiley, New York.

Cicerone, R.J., and R.S. Oremland. 1988. Biogeochemical aspects of atmospheric methane. Global Biogeochem. Cycles 2:299–327.

Conway, T.J., P. Tans, L.S. Waterman, K.W. Thoning, K.A. Masarie, and R.H. Gammon. 1988. Atmospheric carbon dioxide measurements in the remote global troposphere, 1981–1984. Tellus 40B:81–115.

Crutzen, P.J. 1988. Variability in atmospheric-chemical systems. p. 81–108. *In* T. Rosswall et al. (ed.) Scales and global change. John Wiley, New York.

Detwiler, R.P., and C.A.S. Hall. 1988. Tropical forests and the global carbon cycle. Science (Washington, DC) 239:42–47.

Duxbury, J.M., D.R. Bouldin, R.E. Terry, and R.L. Tate III. 1982. Emissions of nitrous oxide from soils. Nature (London) 298:462–464.

Eichner, M.J. 1990. Nitrous oxide emissions from fertilized soils: Summary of available data. J. Environ. Qual. 19:272–280.

Elkins, J.W., and R. Rossen. 1989. Summary report 1988: Geophysical monitoring for climate change, NOAA Environ. Res. Lab., Boulder, CO.

Esser, G. 1987. Sensitivity of global carbon pools and fluxes to human and potential climatic impacts. Tellus 39B:245–260.

Esser, G. 1990. Modelling global terrestrial sources and sinks of CO_2 with special reference to soil organic matter. p. 247–261. *In* A.F. Bouwman (ed.) Soils and the greenhouse effect. John Wiley, New York.

Folland, C.K., T.R. Karl, and K.Y.A. Vinnikov. 1990. Observed climate variations and change. p. 201–238. *In* J.T. Houghton et al. (ed.) Climate change: The IPCC scientific assessment. Cambridge Univ. Press, Cambridge.

Galbally, I.E. 1984. Factors controlling NO_x emissions from soils. p. 23–27. *In* M.O. Andreae and B.S. Schimel (ed.) Exchange of trace gases between terrestrial ecosystems and the atmosphere. John Wiley, New York.

Goudriaan, J., and D. Ketner. 1984. A simulation study for the global carbon cycle, including man's impact on the biosphere. Clim. Change 6:167–192.

Hansen, J., A. Lacis, and M. Prather. 1989. Greenhouse effect of chlorofluorocarbons and other trace gases. J. Geophys. Res. 94:16 417–16 421.

Houghton, R.A., R.D. Boone, J.R. Fruci, J.E. Hobbie, J.M. Mellilo, C.A. Palm, B.J. Peterson, G.R. Shaver, G.M. Woodwell, B. Moore, D.L. Skole, and N. Myers. 1987. The flux of carbon from terrestrial ecosystems to the atmosphere in 1980 due to changes in land use: Geographical distribution of the global flux. Tellus 39B:122–139.

Houghton, R.A., J.E. Hobbie, J.M. Melillo, B. More, B.J. Peterson, G.R. Shaver, and G.M. Woodwell. 1983. Changes in the carbon content of terrestrial biota and soils between 1860 and 1980: A net release of CO_2 to the atmosphere. Ecol. Monogr. 53:235–262.

Houghton, R.A., and D.L. Skole. 1990. Changes in the global carbon cycle between 1700 and 1985. *In* B.L. Turner (ed.) The earth transformed by human action. Cambridge Univ. Press, Cambridge. (In press.)

Intergovernmental Panel on Climate Change. 1990. Introduction. p. xxxvi–xxxvii. *In* J.T. Houghton et al. (ed.) Climate change: The IPCC scientific assessment. Cambridge Univ. Press, Cambridge.

Keeling, C.D. 1986. Atmospheric CO_2 concentrations—Mauna Loa observatory, Hawaii 1958–1986. NDP-001/RI Carbon Dioxide Information Center, Oak Ridge Natl. Lab., Oak Ridge, TN.

Keller, M., W.A. Kaplan, and S.C. Wofsy. 1986. Emissions of N_2O, CH_4, and CO_2 from tropical forest soils. J. Geophys. Res. 91:11 791-11 802.

Kern, J.S., and M.G. Johnson. 1991. The impact of conservation tillage use on soil and atmospheric carbon in the contiguous United States. EPA/600/3-91-056. USEPA, Environ. Res. Lab., Corvallis, OR.

Khalil, M.A.K., and R.A. Rasmussen. 1984. Global sources, lifetimes, and mass balances of carbonyl sulfide (COS) and carbon disulfide (CS_2) in the Earth's atmosphere. Atmos. Environ. 18:1805-1813.

Lacis, A.A., D.J. Wuebbles, and J.A. Logan. 1990. Radiative forcing of global climate changes in the vertical distribution of ozone. J. Geophys. Res. 95:9971-9981.

Livingston, G.P., P.M. Vitousek, and P.A. Matson. 1988. Nitrous oxide flux and nitrogen transformations across a landscape gradient in Amazonia. J. Geophys. Res. 93:1593-1599.

Matson, P.A., and P.M. Vitousek. 1987. Cross-system comparisons of soil nitrogen transformations and nitrous oxide flux in tropical forest ecosystems. Global Biogeochem. Cycles 1:163-170.

Mayewsky, P.A., W.B. Lyons, M. Twickler, W. Dansgaard, B. Koci, C.I. Davidson, and R.E. Honrath. 1986. Sulfite and nitrate concentrations from a South Greenland ice core. Science (Washington, DC) 232:975-977.

Melillo, J.M., and P.A. Steudler. 1989. The effect of nitrogen fertilization on the COS and CS_2 emission from temperate forest soils. J. Atmos. Chem. 9:411-417.

Minami, K., and K.S. Fukushi. 1981a. Volatilization of carbonyl sulfide from paddy soils treated with sulfur containing substances. Soil Sci. Plant Nutr. 27:389-345.

Minami, K., and K.S. Fukushi. 1981b. Detection of carbonyl sulfide among gases produced by the decomposition of cystine in paddy soils. Soil Sci. Plant Nutr. 27:105-109.

Mitchell, J.F.B., S. Manabe, V. Meleshko, and T. Tokioka. 1990. Equilibrium climate change—and its implications for the future. p. 137-164. In J.T. Houghton et al. (ed.) Climate change: The IPCC Scientific Assessment. Cambridge Univ. Press, Cambridge.

Mooney, H.A., P.M. Vitousek, and P.A. Matson. 1987. Exchange of materials between terrestrial ecosystems and the atmosphere. Science (Washington, DC) 238:926-932.

Mosier, A., D. Schimel, D. Valentine, K. Bronson, and W. Parton. 1991. Methane and nitrous oxide fluxes in native, fertilized and cultivated grasslands. Nature (London) 350:330-332.

Mosier, A.R., and D.S. Schimel. 1991. Influence of agricultural nitrogen on atmospheric methane and nitrous oxide. Chem. Ind. (London) 23:874-877.

Neftel, A., A.J. Beer, H. Oeschger, F. Zurcher, and R.C. Finkel. 1985. Sulfate and nitrate concentrations in snow from South Greenland 1895-1979. Nature (London) 314:611-613.

Ojima, D.S., A.R. Mosier, W.J. Parton, D.S. Schimel, and D.W. Valentine. 1992. Effect of land use change on soil methane oxidation. Chemosphere. (In press.)

Schlesinger, W.H. 1990. Biogeochemistry: An analysis of global change. Acad. Press, San Diego.

Servant, J. 1989. Les sources et les puits d'oxysulfure de carbone (COS) a l'echelle mondiale. (In French.) Atmos. Res. 23:105-116.

Shine, K.P., R.G. Derwent, D.J. Wuebbles, and J-J. Morcrette. 1990. Radiative forcing of climate. p. 47-68. In J.T. Houghton et al. (ed.) Climate change: The IPCC scientific assessment. Cambridge Univ. Press, Cambridge.

Steele, L.P., P.J. Fraser, R.A. Rasmussen, M.A.K. Khalil, T.J. Conway, A.J. Crawford, R.H. Gammon, K.A. Masarie, and K.W. Thoning. 1987. The global distribution of methane in the troposphere. J. Atmos. Chem. 5:125-171.

Steudler, P.A., R.D. Bowden, J.M. Melillo, and J.D. Aber. 1989. Influence of nitrogen fertilization on methane uptake in temperate soils. Nature (London) 341:314-316.

Steudler, P.A., and B.J. Peterson. 1984. Contribution of gaseous sulfur from salt marshes to the global sulfur cycle. Nature (London) 311:455-457.

Tans, P.P., T.J. Conway, and T. Nakasawa. 1989. Latitudinal distribution of sources and sinks of atmospheric carbon dioxide. J. Geophys. Res. 94:5151-5172.

Tans, P.P., I.Y. Fung, and T. Takahashi. 1990. Observational constraints on the global atmospheric carbon dioxide budget. Science (Washington, DC) 247:1431-1438.

U.S. Environmental Protection Agency. 1989. Policy options for stabilizing global climate. p. 20. In D.A. Lashof and D.A. Tirpak (ed.) Draft report to Congress. USEPA Office of Policy, Planning and Evaluation, Washington, DC.

U.S. Environmental Protection Agency. 1990. Greenhouse gas emissions from agricultural systems. Vol. 1, Summary report. p. 19-24. In Proc. of IPCC-RSWG Workshop, Washington, DC. 12-14 Dec. 1989. USEPA Climate Change Div., Washington, DC.

Vaghjiani, G.L., and A.R. Ravishankara. 1991. New measurements of the rate coefficient for the reaction of OH with methane. Nature (London) 350:406–408.

Watson, R.T., H. Rodhe, H. Oeschger, and U. Siegenthaler. 1990. Greenhouse gases and aerosols. p. 7–40. *In* J.T. Houghton et al. (ed.) Climate change: The IPCC scientific assessment. Cambridge Univ. Press, Cambridge.

Whalen, S.C., and W.S. Reeburgh. 1990. Consumption of atmosphere methane by tundra soils. Nature (London) 346:160–162.

Wiersum, K.F., and P. Ketner. 1989. Reforestation, a feasible contribution to reducing the atmospheric carbon dioxide content. *In* P.O. Okken et al. (ed.) Climate and energy: The feasibility of controlling CO_2 emissions. Kluwer, Doordrecht, the Netherlands.

2 Methods for Measuring Atmospheric Gas Transport in Agricultural and Forest Systems

O.T. Denmead and M.R. Raupach

CSIRO Centre for Environmental Mechanics
Canberra, Australia

Our task is to present an overview of methods for measuring exchanges of greenhouse gases between agricultural and forest systems and the atmosphere. Most of the paper is concerned with the gases N_2O, CH_4, and CO_2, whose increasing concentrations account for more than 70% of the predicted atmospheric warming referred to loosely as the greenhouse effect. The overview applies equally as well to other trace gases with sources and sinks in the terrestrial biosphere, such as NH_3, O_3, NO_x, SO_2, and terpenes, however.

Other papers in this volume outline the biogeochemistry of the three main greenhouse gases. Suffice it to say that all of them have significant sources and/or sinks in the soil–plant system. Perhaps the biggest uncertainty in our understanding of the biogeochemistry of these gases is our knowledge of their source and sink strengths. In turn, much of this uncertainty derives from deficiencies in our ability to measure their surface to air exchange rates and to extrapolate those measurements to regions, land types and the globe.

The techniques commonly employed to measure gas fluxes in the field have been reviewed by a number of authors recently: chambers by Mosier (1989), Schütz and Seiler (1989), Wesely et al. (1989), and Denmead (1991); land-based micrometeorological techniques by Baldocchi et al. (1988), Fowler and Duyzer (1989), Wesely et al. (1989), and Denmead (1991); and aircraft measurements by Desjardins and MacPherson (1989) and Wesely et al. (1989). Chamber techniques are discussed again in this volume by Hutchinson and Livingston (1993), and micrometeorological techniques, including aircraft measurements, by Desjardins et al. (1993). Accordingly, chamber and conventional land-based micrometeorological techniques are discussed briefly here, and aircraft techniques not at all. Space is devoted to some unconventional micrometeorological techniques that offer considerable promise for measuring gas exchange on horizontal scales ranging from meters to kilometers.

MEASUREMENT TECHNIQUES

General

While micrometeorological techniques will usually be preferred for measuring trace gas fluxes because they do not disturb the sample area or its microclimate, there are circumstances in which they are neither feasible nor desirable, and enclosures offer the only practicable approach. The biggest limitation on the use of micrometeorological techniques is usually the lack of sensors that combine sensitivity, fast response, continuous operation, ruggedness, and low cost. The problem is worst for N_2O and CH_4 for which chambers have been the only feasible technique until quite recently. In the last 4 years, new gas detection systems based on lasers have been developed for both gases. These make micrometeorological approaches possible. The sensors are expensive (approximately $100 000) and are likely to be under development and require specialist operators for some time yet, however. Nonetheless, continuing instrument development is likely to make micrometeorological measurements of N_2O and CH_4 routine in the near future.

Even so, there will doubtless be a continuing need to employ chambers for measuring trace gas fluxes because of their portability, the small land areas they require, and their versatility. They permit process studies and experiments with land treatments in numbers that cannot be contemplated with micrometeorological approaches because of the large land areas that the latter require.

Chambers

The principle is to restrict exchange of air with the atmosphere so that changes in the concentration of the emitted (or absorbed) gas in the head space can be detected readily. Sensor sensitivity is not always the reason for using chambers, however. The need for experiments with many treatments has been mentioned already. Another is the unsuitability of some sensing equipment for field use. Very small differences in gas concentration can be resolved by gas chromatographs and mass spectrometers for instance, but these are usually relatively slow-response, laboratory instruments, unsuitable for continuous monitoring or deployment in the field.

Chamber systems fall into two categories: closed systems in which there is no replacement of air in the head space and the gas concentration changes continuously, and open systems where a constant flow of air through the head space is maintained and the gas concentration attains a steady difference from the background concentration in the ambient air. The former system is used more commonly because a larger concentration change occurs and the system is mechanically simpler. (No power may be required for pumps or for on-line monitors, for instance, syringe samples for later analysis may be sufficient.) In closed systems, the rate of concentration change is used to calculate the gas flux to or from the soil system

$$F_g = (V/A) \, d\rho_g/dt \qquad [1]$$

where F_g is the flux density of the gas, V is the volume of the head space, A is the area of land enclosed, ρ_g is the gas density within the chamber, and t is time. Of course, precautions must be taken to prevent changes in the gas concentration within the chamber from influencing the flux. This aspect is dealt with in detail by Hutchinson and Livingston (1993). In open systems, the flux is calculated from

$$F_g = v \, (\rho_g - \rho_b)/A \qquad [2]$$

where v is the gas flow rate and ρ_b is the background gas concentration. In practice, v is adjusted so as to keep $(\rho_g - \rho_b)$ small.

In closed systems where air may be circulating in a loop through a monitor, and in open systems where air is pumped through the head-space, the pressure differential in the chamber, induced by pumping, should be monitored so as to keep it negligibly small. Very small pressure differentials can induce a mass flow of air into or out of the chamber through the soil (Denmead, 1979a). Jury et al. (1982) have examined problems of inference from both closed and open systems from a theoretical standpoint. They draw attention to typical diffusion times in the soil profile, the possibilities of gas production and consumption during diffusion, how these affect the concentration profiles developed in the soil underneath the chamber, and what the flux of gas into the chamber represents. They point out that when production occurs deep within the soil, the appropriate sampling period for a production event may be many days. If sampling over several days is required, then open systems that have minimal effects on soil concentration profiles seem to be the only useful option.

Another aspect of chamber operation is of relevance here: the influence of the chamber's microclimate on the gas exchange of soils or plants within it. The production of N_2O, CH_4 and CO_2 in soil result from biological activity and like most biological processes, production is strongly temperature dependent. Denmead et al. (1979) give a Q_{10} (relative change in activity for a $10\,°C$ change in temperature) of 2.8 for N_2O production in the field. Conrad (1989) gives values of 2.5 to 3.5 for CH_4 and Monteith et al. (1964) give a value of 3 for CO_2. Large diurnal variations in soil emissions of these gases are thus to be expected, and maintaining soil temperatures inside the chamber close to those outside is important. Soil temperature is governed by energy partitioning at the soil surface, but maintaining the natural energy balance of the outdoors inside a chamber is very difficult. If gas flux measurements are to extend over several days, it may be wiser to use several short enclosure periods than to measure continuously at the same site.

Given the temperature dependence of many trace gas emissions, and uncertain phase relationships resulting from differing production zones and differing diffusion properties in the soil profile (Blackmer et al., 1982; Jury et al., 1982), the minimum period for flux measurement would seem to be 24 h. Spot measurements, although often published, are usually quite misleading; the practice of reporting spot measurements in "annual" units like "kg ha $^{-1}$ year $^{-1}$" can only be deplored. Furthermore, some species, such as N_2O and CO_2, emit episodically, depending on soil water supply. The minimum measuring period might then be an entire drying cycle.

Nitrous oxide exchange is predominantly between soil and atmosphere, but most of the exchange of CO_2, and of CH_4 in wetland–plant systems is between plants and atmosphere (Seiler et al., 1984; Nouchi et al., 1990; Rennenberg, 1992). In addition, we might expect that CH_4 production in the rhizosphere of wetland plants will depend partly on their CO_2 assimilation rate. Hence for the C gases at least, realistic measurements of gas exchange in chambers over periods of more than a few minutes require preservation of those environmental conditions that influence plant functioning and CO_2 exchange in the out-of-doors: ventilation, atmospheric CO_2 concentrations, evaporation rate, light, temperature, humidity, and so on.

Finally, we observe that the surface area enclosed by chambers is relatively small, often < 1 m^2, whereas the point to point variability of soil gas emissions is notoriously large. Variations in soil emission rates by factors of two within distances of a few meters are commonplace and 10-to-1 variations occur frequently, e.g., Matthias et al. (1980), Galbally et al. (1985). Therefore, chamber flux measurements tend to be highly variable. The study of spatial variability of field-measured denitrification gas fluxes by Folorunso and Rolston (1984) provides a concrete example. They found coefficients of variation of 282 to 379% in chamber measurements of N_2O flux. Using geostatistics, they calculated that it would require as many as 350 measurements to estimate the true mean flux within $\pm 10\%$ on a 3- $\times$ 36-m experimental plot. Yet our knowledge of global soil emissions of N_2O is based almost exclusively on a relatively small number of chamber investigations, many of them making only spot measurements. The belief that tropical forest soils are the largest global source of N_2O is based on < 1000 spot chamber flux measurements. One of the first estimates of global emissions from this source (Keller et al., 1986) was based on 61 such measurements, each of 20- to 40-min duration. The situation for CH_4 is not much better, although there is at least some corroborating evidence for the relative sizes of the terrestrial CH_4 sources from atmospheric isotope concentrations (Wahlen et al., 1989). We note, for instance, that all the evidence identifying rice (*Oryza sativa* L.) paddies as the most important anthropogenic source of atmospheric CH_4 (Watson et al., 1990) comes from chambers. There is a pressing need for nondisturbing measurements of trace gas fluxes, which integrate over larger areas and longer time scales. Micrometeorological techniques offer more promise in these respects.

MICROMETEOROLOGICAL METHODS

Gradient Techniques

These are based on the proposition that in the unobstructed air layer above the surface of interest, gas transport occurs by a process of turbulent diffusion along the gradient of mean concentration. Practical applications of these and eddy correlation techniques are limited to situations in which the air has blown over a homogeneous exchange surface for a long distance so that profiles of gas concentration in the air are in equilibrium with the local rate of exchange, and horizontal concentration gradients are negligible. In such circumstances, the vertical flux density of the gas will be constant with height in the air layers close to the ground and a one-dimensional (vertical) analysis can be made.

It is conventional to define the upper boundary of the air layer whose gas concentration is modified by an emission at the ground as the height, Z, at which the increase is reduced to some negligibly small fraction, usually < 10%, of its highest value in the layer (Sutton, 1953). As a rough guide, Z is about one-tenth of the fetch, i.e., the distance the wind has travelled over the treated area. The depth of the air layer in which the flux is constant with height is much smaller; suggested fetch/height ratios vary from about 200:1 (Dyer, 1963) to 100:1 (Wesely et al., 1982). For a fetch of 100 m, these ratios would give maximum working heights of only 0.5 to 1 m. Many eddy correlation instruments, however, can only be operated successfully at heights of 2 m or more above the surface because of limitations to their frequency response. Likewise, the small size of atmospheric gradients often requires that concentration differences be measured over a height interval of 2 or 3 m. Even with the less stringent fetch–height ratio of 100:1, it is evident that experimental areas need to be hundreds of meters in extent.

Following Webb et al. (1980), we can write for the vertical flux density of the gas

$$F_g = - \overline{\rho_a} K_g \, \partial \overline{s} / \partial z \qquad [3]$$

where ρ_a is the density of dry air, K_g is an eddy diffusivity, $s(= \rho_g / \rho_a)$ is the mixing ratio of the gas with respect to dry air, z is height, and the overbar denotes a time average. The eddy diffusivity depends on wind speed, surface roughness, height above the surface, and atmospheric stability. It is usual to infer its value in situ from flux–gradient relationships like Eq. [3], written for some other tracer entity whose flux is known and whose concentration gradient is measured simultaneously. Heat, water vapor, and momentum are the tracers used most commonly. The assumption is generally made that

$$K_H = K_V = K_g, \qquad [4]$$

i.e., the transport mechanisms are similar for scalar fluxes. In Eq. [4], subscripts H and V refer to heat and water vapor respectively.

The relationship between the eddy diffusivity for momentum K_M and K_g depends on atmospheric stability. If momentum is used as the tracer, additional measurements of temperature stratification are required to quantify the stability effect. Appropriate procedures and formulations for these various approaches are contained in many of the publications referred to in the Introduction. Denmead (1983) describes their application to measurement of fluxes of trace gases generally.

Focusing on the trace gases of interest here, we note that gradient techniques were developed for measuring CO_2 fluxes more than 30 yr ago (Inoue et al., 1958; Lemon, 1960; Monteith & Szeicz, 1960), and are now used commonly. With good technique, differences in atmospheric CO_2 concentration smaller than 0.1 ppm_v can be detected by on-line infrared gas analysis, a resolution that is sufficient for most applications in agricultural and forest systems. The situation for N_2O and CH_4 is not as satisfactory because of sensor limitations. Gradient methods have been used to measure N_2O fluxes from agricultural fields by Lemon (1978) and Hutchinson and Mosier (1979), but the fluxes were exceptionally large (mostly > 1000 ng m^{-2} s^{-1}). Denmead (1979a) calculated that for the flux densities of N_2O more common in agriculture and forestry, which are generally one to two orders of magnitude smaller, resolutions of tenths of a part per billion will be required to define the vertical profile of atmospheric N_2O concentration accurately. Such resolution appears to be still on the edge of what can be achieved with gas chromatography and beyond what can be accomplished with on-line infrared gas analysis. It appears, however, that this resolution can now be achieved with laser systems (G.W. Thurtell, 1991, personal communication).

For CH_4, we are aware of only one reported application of gradient methods in an agricultural system, a preliminary account by Denmead (1991) of measurements of emissions from a rice field, which employed on-line infrared gas analysis to measure the concentration gradient and measurements of the profile of horizontal wind speed to infer K_g. Figure 2–1 is taken from that paper. As pointed out there, the success of the method hinges on how accurately the CH_4 gradient can be determined. The magnitude of the latter depends on the emission rate and the wind speed. By day, with typical wind speeds of 4 m s^{-1}, gradients over the rice paddy were usually between 10^{-3} and 10^{-2} ppm_v m^{-1}, and were sometimes difficult to distinguish from the system noise. By night, with lower wind speeds and apparently larger fluxes, gradients were almost an order of magnitude greater, but wind speeds were more difficult to determine accurately (Fig. 2–1a,b). Despite these difficulties, the approach is a feasible alternative to chambers and refinements in technique promise to make it a very useful tool for making nondisturbing, continuous measurements of CH_4 emissions on horizontal scales of 100 m or more. In this instance, the magnitudes of the emissions and their diurnal pattern (Fig. 2–1c) were remarkably similar to those measured elsewhere with chambers, including the occurrence of a nighttime maximum lagging behind water temperature (Fig. 2–1d), as observed by Seiler et al. (1984). As for N_2O, laser systems appear to offer a more accurate measurement of the atmospheric CH_4 gradient.

To conclude this section, it should be said that even when the fetch requirements discussed earlier in the paper are met, it is unrealistic to expect micrometeorological methods to provide reliable flux measurements 24 hours a day. Rain is an obvious hazard and measurements at night usually are difficult: anemometers become unreliable at low winds, net radiation is small and difficult to measure with precision, dew collects on radiometers, boundary layers are often not well developed, and rapid changes in the stratification of the boundary layer can make time-averaged concentration profiles unreliable. Some of these comments also apply to periods of low radiation or low wind speed during the daytime. In a later section, we describe a convec-

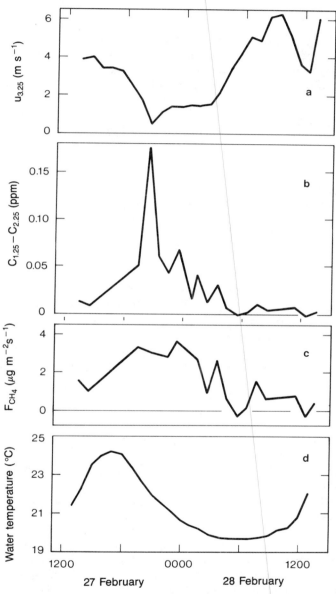

Fig. 2–1. Diurnal courses of: (a) wind speed at 3.25 m above water surface in rice field at Griffith, NSW, (b) corresponding differences in CH_4 concentration between heights of 1.25 and 2.25 m, (c) calculated flux of CH_4 from rice field, (d) temperature of floodwater (from Denmead, 1991).

tive boundary layer budget method that avoids many of these difficulties, and indeed relies for its success on changes in the attributes of the boundary layer between night and day.

Eddy Correlation

While often regarded as the best micrometeorological option because it is a direct measurement requiring no assumptions about eddy diffusivities, stability corrections or the shape of the wind profile, eddy correlation has its own difficulties when applied to some trace gases. These are outlined below.

The instantaneous vertical flux density of a gas is simply the product of the vertical wind speed w and the gas density ρ_g. The mean flux density F_g is given by

$$F_g = \overline{w\rho_g} \qquad [5]$$

where the bar denotes an average over an appropriate measuring period. Both w and ρ_g can be represented as sums of means, $\overline{w}$ and $\bar{\rho}_g$ and fluctuations about those means, w' and ρ_g'. Thus

$$F_g = \overline{w}\,\bar{\rho}_g + \overline{w'\rho_g'} \qquad [6]$$

It has been common to assume that over a uniform, horizontal surface $\overline{w} = 0$ and hence $\overline{w}\,\bar{\rho}_g = 0$, but Webb et al. (1980) argue that $\overline{w}$ will be non-zero whenever there is a flux of heat and/or water vapor between the surface and the atmosphere. Since $\overline{w}$ will generally be <1 mm s^{-1}, it cannot be measured directly, but its magnitude can be calculated if the simultaneous flux densities of heat H and water vapor E are known. Webb et al. (1980) show that Eq. [6] then becomes

$$F_g = \overline{w'\rho_g'} + (\bar{\rho}_g/\bar{\rho}_a)\,[\mu/\,(1 + \mu\sigma)]\,E + (\bar{\rho}_g/\bar{\rho})H/c_p\,\overline{T} \qquad [7]$$

In Eq. [7] $\mu = m_a/m_v$; m_a is the "molecular weight" of dry air and m_v is the molecular weight of water vapor; $\sigma = \bar{\rho}_v/\bar{\rho}_a$; $\bar{\rho}_v$ is the density of water vapor; the total air density $\rho = \rho_a + \rho_v$; c_p is the specific heat of air at constant pressure; and T (K) is air temperature.

The first term on the right-hand side of Eq. [7] is the apparent eddy flux calculated on the assumption that $\overline{w} = 0$. The other two terms are the corrections for the density effects due to water vapor and heat transfer. As Eq. [6] indicates, their importance depends on the magnitude of $\bar{\rho}_g$. For trace gases with small background concentrations and relatively large fluxes, such as ammonia, they are unimportant, but for all of N_2O, CH_4, and CO_2 they will usually be very important. Wesely et al. (1989) tabulate typical corrections, based on published estimates of the gas fluxes and H and λE (λ being the latent heat of evaporation). For N_2O, the ratio $\overline{w}\,\bar{\rho}_g/F_g$ varies from 0.25 to 10.5 and for CH_4 from 0.04 to 0.56. For CO_2, Wesely et al. (1989) give one value for low CO_2 fluxes of 0.76. For productive crops with high rates of photosynthesis, $\overline{w}\,\bar{\rho}_g/F_g$ is typically about 0.3.

The magnitude of the corrections will usually mean that they cannot be ignored, and additional measurements of H and E will be required to evaluate them. In turn, this introduces further uncertainty in the measurement because the true flux will often be the difference between two large, experimentally determined numbers. Denmead and Bradley (1989) and Leuning and King (1992) describe eddy-correlation systems in which the air was pumped down a tube to a closed-path analyzer for measurement of the gas (CO_2) concentration. In those cases, sampling down the tube minimized the temperature fluctuations and so the need for the correction term in H. Because the fluctuations in gas concentrations are also damped by sampling down the tube, a further set of corrections may be necessary to account for the damping, however. The problem is discussed in Lenschow and Raupach (1991) and Leuning and King (1992).

Leuning and Moncrieff (1990) describe yet another correction in eddy-correlation measurements of the CO_2 flux, that due to cross-sensitivity of the analyzer to water vapor. Its magnitude varies with individual instruments and the flux densities of heat and water vapor, but it is typically about 10% of the true CO_2 flux.

Webb et al. (1980) point out that if gas concentrations are measured relative to dry air and at a common temperature, no corrections for the density fluctuations are necessary. Usually, this will not be entirely possible in eddy-correlation systems, particularly where gas concentrations are measured in situ with an open-path analyzer. But preconditioning the air streams to remove water vapor and attain constant temperature is almost always possible when fluxes are calculated by gradient methods. For a gas like N_2O where the corrections can be so large that the true flux may be only one-tenth of that indicated by in situ eddy correlation, it may be preferable to use a gradient method as a matter of course, even though the sensor has a fast response.

Finally, we note that the situation regarding gas sensors is the same for eddy correlation as for gradient methods: adequate fast-response sensors for CO_2 are available commercially (both open- and closed-path), but fast-response sensors for N_2O and CH_4 are still under development.

Eddy Accumulation

The technique is described in this volume by Desjardins et al. (1993). Theoretical and practical difficulties exist (Hicks & McMillen, 1984), but these are being overcome and it seems likely that eddy accumulation or a variant, conditional sampling (Businger & Oncley, 1990), will eventually be used widely for measuring trace gas fluxes. Instead of a fast-response sensor, these techniques use fast sampling. Air associated with updrafts and downdrafts is sampled into each of two bins at a rate proportional to the vertical velocity. Slow-response, but high-resolution sensors can then be used to measure the mean concentrations of the gas in the "up" and "down" bins at the end of a sampling period. The resolution required in measuring gas concentrations will be much the same as for eddy correlation or gradient methods, but the measuring instrument can be remote and operated under ideal laboratory conditions. The system has another advantage over in situ eddy correlation: the air samples can be preconditioned to avoid the corrections associated with density fluctuations.

Mass Balance

When discussing chambers, we pointed to the very large spatial and temporal variability usually found in gas fluxes from soils and suggested that there was a need for nondisturbing techniques that integrate over larger space and time scales than is possible with chambers. Denmead et al. (1977), Beauchamp et al. (1978), Wilson et al. (1982, 1983), and Denmead (1983) describe a small-plot, micrometeorological, mass balance technique that meets these criteria, but does not require the large fetches needed for gradient and eddy-correlation approaches. Plot dimensions are typically tens of meters instead of the hundreds of meters required for the last. As well, the instrumentation requirements can be quite simple, particularly when trapping techniques can be used to concentrate the gas, as for NH_3 and N_2O.

The method equates the flux of gas into the atmosphere from a treated area of limited upwind extent with the rate at which it is transported by the wind across the downwind edge. The conservation of mass gives

$$F_g = \frac{1}{X} \int_0^Z \overline{u(\rho_g - \rho_b)} \, dz \qquad [8]$$

where X is the upwind fetch, u is horizontal wind speed, and as before, ρ_b is the upwind, background concentration. Here, the aim is to measure gas concentration profiles right through the boundary layer developed over the plot rather than just in the much smaller, equilibrium layer close to the surface. Unlike the micrometeorological methods described so far, this technique is most successful when the fetch is small so that Z is also conveniently small. When X is 10 to 20 m, Z is around 1 to 2 m, its actual magnitude depending on fetch, surface roughness and atmospheric stability.

One restriction on this approach is that the effective gas concentration is the concentration in excess of background. Not only must the upwind concentration profile be measured as well as the downwind, but calculation of the flux through Eq. [8] also requires subtraction of experimentally determined data, which can be an error-prone procedure. The technique is thus suited best to investigations of trace gas fluxes in situations where ρ_b is small and experimental treatments generate fluxes that are large compared with normal emissions. One such situation where the method has been used extensively is measurement of NH_3 emissions following application of N amendments to soils, e.g., Beauchamp et al. (1978), Wilson et al. (1983), Freney et al. (1992).

A problem with practical evaluation of Eq. [8] is that the term $\overline{u\rho_g}$ is the mean of instantaneous fluxes

$$\overline{u\rho_g} = \bar{u}\bar{\rho}_g + \overline{u'\rho_g'}.$$ [9]

The first term on the right of Eq. [9] represents the mass flux out of the treated area due to the mean flow of the wind. The second term represents a turbulent diffusive flux in the opposite direction. The first term, the product of the mean wind speed and the mean concentration, is the term that will be measured usually. Field tests by Leuning et al. (1985) suggest that it over-estimates the true mass flux $\overline{u\rho_g}$ by about 15%. Wind tunnel experiments by Raupach and Legg (1984) suggest 10%, while theoretical calculations by Wilson and Shum (1992) suggest 20%. Probably, it will be enough to apply an empirical correction of, say 15%, although bigger corrections may be need-ed in very small plots and where the surface roughness is large (Wilson & Shum, 1992).

A difficulty of a similar kind is that the fetch X must be known precise-ly in order to calculate F_g. If the experimental area is the usual agronomist's rectilinear plot, X will vary with wind direction and will need to be deter-mined frequently. As discussed by Wilson et al. (1982) and Denmead (1983), this complication can be overcome by working with a circular plot and mea-suring $\bar{u}$ and $\bar{\rho}_g$ at its center. Regardless of compass direction, the wind will always blow towards the center and X will always be equal to the plot radius. Circular plots with radii between 3.5 m (Gordon et al., 1988) and 36 m (Beauchamp et al., 1978) have been employed.

Denmead (1983) points out that apart from its eminent suitability for small plots, this method has some distinct advantages over the equilibrium methods discussed previously. First, it has a simple theoretical basis: it re-quires no special form for the wind profile and no corrections for thermal stratification. Second, if we are prepared to apply an empirical correction to account for the slight overestimation of F_g brought about by ignoring horizontal diffusion, the instrumentation is relatively simple. It needs neither to be of fast response, nor to have the high precision required for gradient measurements.

A further advantage is that both theory and experiment indicate that in certain situations, it is possible to infer the surface flux from measure-ments of the horizontal flux at just one height above the plot center. There are two requirements: although the treated plots should be small, the area in which they are located should be large and uniform so that the wind pro-files are equilibrium profiles. Second, the treated area should have no vegeta-tive cover, or at least a very short one, so that virtually all of the horizontal flux occurs in the unobstructed air layer above the surface. Then, the profile of horizontal flux density has a theoretically predictable shape that is deter-mined by surface roughness, plot geometry and atmospheric stability (Philip, 1959; Mulhearn, 1977; Wilson et al., 1982). From their analysis of the in-fluence of stability on profile shape, Wilson et al. (1982) predict the exis-tence of a particular height within the plot boundary layer, at which the normalized horizontal flux, $\bar{u}\bar{\rho}_g/F$, has almost the same value in all stability regimes. They call that height ZINST. If the appropriate value of $\bar{u}\bar{\rho}_g/F$ at ZINST is known, measurements of $\bar{u}$ and $\bar{\rho}_g$ at only that height are sufficient to determine F_g.

Experimental and theoretical evidence for the existence of a ZINST and methods for its determination are given by Wilson et al. (1982, 1983) and Denmead (1983). The concept has been exploited in many experiments conducted by the senior author and his colleagues, e.g., Freney et al. (1992), who have used a passive sampler described by Leuning et al. (1985) to measure the horizontal flux of NH_3 at ZINST, and hence infer the surface flux, in circular plots of 25-m radius. The same general system, with either full profile measurements or measurements at ZINST only, could be exploited for other gases such as N_2O and CH_4, although the higher background concentrations of these gases (relative to ρ_g) might make application more difficult than for NH_3. Jarvis (1990) describes an application to measurements of N_2O emissions from grassland soils.

Lagrangian Methods

These methods, recently developed by Raupach (1989a, b), offer a relatively simple means for inferring fluxes of trace gases and their source–sink distributions within plant canopies. They require measurements of the mean gas concentration profiles within the canopy and some knowledge of turbulence and Lagrangian time scales in that space. As for chambers, advantage is taken of the restricted air movement in the canopy, which makes for larger and more easily measurable concentration changes than in the unobstructed atmosphere.

Within the canopy, the source or sink strength of the gas is designated by $S(z)$. It is related to the vertical flux density at any level by

$$S(z) = dF_g(z)/dz, \qquad [10]$$

so that

$$F_g(h) = F_g(0) + \int_0^h S(z)dz \qquad [11]$$

where $F_g(0)$ is the gas flux density at the ground surface and h is canopy height. The Lagrangian dispersion theory developed by Raupach (1989a) enables a prediction of the gas concentration profile, which Raupach designates by $C(z)$, from knowledge of $S(z)$. The inverse Lagrangian method described by Raupach (1989b) builds on that work to allow the inference of $S(z)$ from $C(z)$. The procedure uses a discretized, linear relationship between source and concentration profiles, in which the coefficients are determined by the statistics of the wind field. These statistics determine the contributions of gas emitted at particular heights to the concentrations developed at any height within the canopy. By summing the emissions from all source layers (denoted by the suffix j), the gas concentration at any height (denoted by suffix i) can be written as

$$C_i - C_R = \Sigma \, D_{ij} \, S_j \, \Delta z_j \qquad [12]$$

where C_R is the concentration at a reference height above the canopy.

The coefficients of the dispersion matrix, the D_{ij}, are calculated from profiles of σ_w, the standard deviation of vertical wind speed, and T_L, the integral Lagrangian time scale within the canopy, following Raupach (1989a). Once these are known, Eq. [12] becomes a set of linear equations that can be solved for the source profile S_j.

Figures 2–2 to 2–5 (from Raupach et al., 1992) show the application of the method to the problem of deducing source and sink strengths for water vapor and CO_2 in the canopy of a wheat (*Triticum aestivum* L.) crop with leaf area index of 3.5 and height of 0.76 m. Figure 2–2 shows the profiles of water vapor and CO_2 volumetric concentration determined from measurements over 30 min at two heights above the canopy and six heights within it.

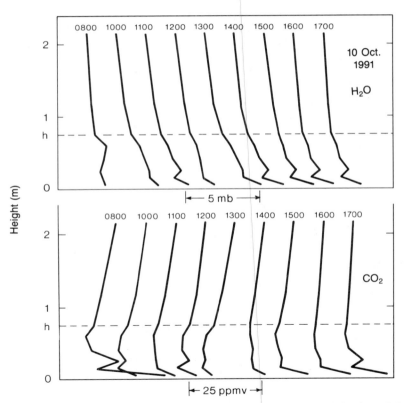

Fig. 2–2. Mean 30-min profiles of water vapor pressure and CO_2 concentration in and above a wheat crop at Wagga Wagga, NSW, for runs commencing at the time indicated. Each profile shows differences from the top measuring height at 2.15 m. The crop height h was 0.76 m.

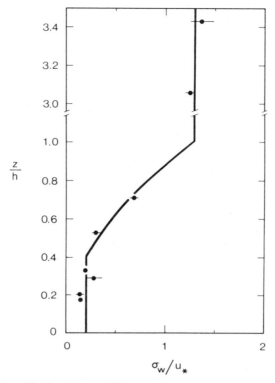

Fig. 2–3. Normalized vertical profile of σ_w in and above the wheat crop of Fig. 2–2.

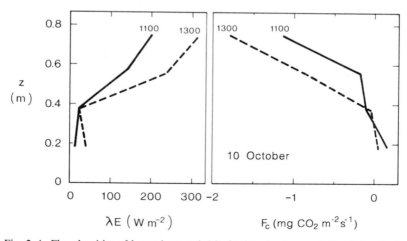

Fig. 2–4. Flux densities of latent heat and CO_2 in the wheat crop of Fig. 2–2, calculated by the inverse Lagrangian method.

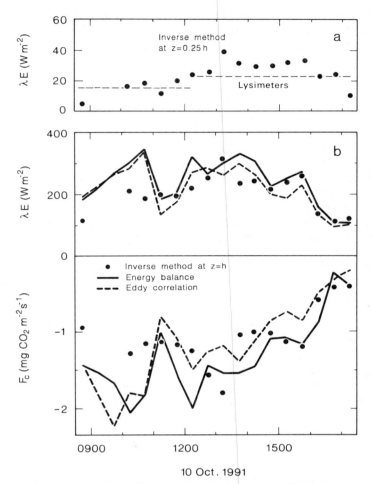

Fig. 2-5. (a) Flux density of latent heat at 0.19 m in wheat crop of Fig. 2-2 calculated by the inverse Lagrangian method (dots), and flux at the soil surface measured by minilysimeters. (b) Flux densities of latent heat and CO_2 at the top of the wheat crop of Fig. 2-2 calculated by the inverse Lagrangian method (dots) and by two conventional micrometeorological methods (lines).

Figure 2-3 shows the normalized profile of σ_w measured with sonic anemometers and used in the calculations. The normalizing factor is the friction velocity u_*. The integral time scale T_L was calculated from a relationship given in Raupach et al. (1992), viz., $T_L u_* /h = \max[0.4(z/h)^{1/2}, 0.1]$. Equation [12] was solved for four source layers, each 0.19 m thick.

Figure 2-4 shows the predicted cumulative fluxes of water vapor and CO_2, obtained from Eq. [11] and [12], on two occasions. For water vapor, the analysis predicts some evaporation at the soil surface and strong contributions from foliage in the top half of the canopy. For CO_2, it predicts some soil respiration and strong assimilation in the top foliage. Figure 2-5a compares the predicted flux of water vapor at 0.19 m with measurements of evaporation at the soil surface obtained by use of minilysimeters (R. Leuning, 1991, personal communication). We should expect the flux at 0.19 m to be close to that at the soil surface but slightly higher because of evaporation from lower leaves, which is, in general, what the analysis predicts. Figure 2–5b compares predictions of the cumulative fluxes at the top of the canopy, obtained from Eq. [11] and [12], with independent measurements of those fluxes in the air layer above the canopy obtained by two "conventional" micrometeorological methods; one method is a gradient approach through the energy balance, the other method is by eddy correlation. Apart from some discrepancies early in the day, the agreement between Lagrangian and conventional methods is excellent, as good as the mutual agreement between the two conventional methods.

This relatively simple computational and measurement scheme obviously has much promise for trace gas flux measurement in plant canopies. Our example shows how useful it can be for gases with sources and/or sinks in the foliage, but it should prove particularly useful for gases like N_2O that have only a soil source. The concentration gradients will be much larger than in the unobstructed air and the analysis within the canopy should be uncomplicated by contributions from the foliage.

Convective Boundary Layer Budget Methods

By day over land, when surface heating generates buoyant convection, the planetary boundary layer consists typically of a progressively deepening layer of strong convective turbulence, often more than 1 km deep by evening, known as the convective boundary layer (CBL). The turbulence mixes the bulk of the CBL very efficiently, producing uniform average (potential) temperatures and concentrations of other scalar entities such as water vapor and CO_2 except in a thin surface layer, a few tens of meters deep. At the top of the CBL, the momentum of ascending thermals causes them to overshoot into the stable overlying air so the well-mixed region extends slightly into the stable air mass and is bounded above by a sharp, capping inversion. The capping inversion height, and hence the CBL depth, increases during the day because of the warming of the entire CBL by daytime solar radiation. This deepening is associated with entrainment of the overlying atmosphere into the CBL. The growth of the CBL during a typical day is shown in Fig. 2–6, from Raupach et al. (1992).

It is well known (and supported by data presented here, Fig. 2–7) that there are substantial diurnal fluctuations of atmospheric CO_2 concentration near the surface of the earth over land. During the night, concentrations build up to large values, typically 100 ppm_v above the ambient (large-scale average) concentration, because the products of respiration are trapped in the shallow, stably stratified, nocturnal boundary layer. By day, photosynthesis at the vegetated surface causes a CO_2 "drawdown," a depression of the CO_2 concentration near the surface below the ambient concentration. The maximum drawdown during the day is typically 20 to 40 ppm_v (Fig. 2–7). Diurnal

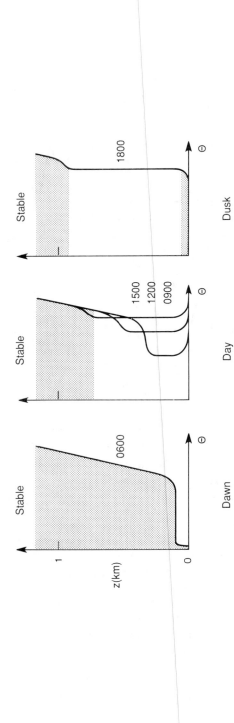

Fig. 2-6. Development of the convective boundary layer during the day. Heavy lines represent profiles of potential temperature, Θ. Stippled areas represent inversion regions. Clear areas represent the well-mixed CBL, (from Raupach et al., 1992).

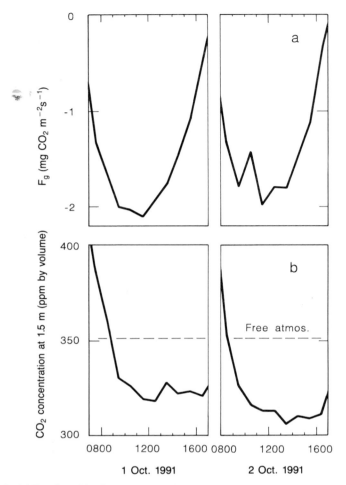

Fig. 2–7. (a) Daytime CO_2 fluxes, and (b) CO_2 drawdown over wheat crop at Wagga Wagga, NSW.

variations in the concentrations of other trace gases are not so well documented, but Denmead (1979b) has reported a distinct diurnal cycle in the ambient N_2O concentration after widespread rain, see below.

Here we consider how measurements of diurnal concentration variations can be used to infer CO_2 (or other trace scalar) fluxes at the surface of the earth on a regional scale (100–1000 km²). This is a budget method for flux measurement that amounts to using the CBL as a large "natural chamber." Its bulk properties are independent of small-scale surface heterogeneities, so it acts as a natural integrator of surface fluxes over heterogenous terrain (Raupach, 1991). The distance over which the CBL carries some "memory" of conditions upwind, its "footprint," is typically 5 to 30 km.

Following Raupach (1991), the conservation equation for scalars in the CBL can be written as

$$\frac{dC}{dt} = \frac{F_c}{L} + \frac{C_+ - C}{L} \cdot \frac{dL}{dt} \qquad [13]$$

where C is the mean scalar concentration in the CBL, t is the time, L is the CBL depth, and C_+ is the value of C just above L. The first term on the right-hand side of Eq. [13] accounts for the change in C due to the surface flux, and the second for the change due to entrainment of air at the top of the CBL. Rearranging Eq. [13] leads to an expression for the regional flux

$$F_c = L \cdot \frac{dC}{dt} - (C_+ - C) \cdot \frac{dL}{dt} \qquad [14]$$

With some simplifying assumptions, Raupach et al. (1992) derive an integrated form of Eq. [13], which allows direct calculation of the total gas flux over periods of several hours, viz.,

$$I_g(t) = -L(t)[C_+ - C(t)] \qquad [15]$$

where I_g is the cumulative gas flux over time t.

To explore the utility of these approaches, we have estimated regional fluxes of CO_2 and N_2O (via Eq. [15] for CO_2 and Eq. [14] for N_2O), using observations of the diurnal variation in their near-surface concentrations (at 1.5 m). In our calculations we have chosen C_+ to be the free-atmosphere, base-line values for CO_2 and N_2O at the times of observation. In each case we have compared the estimated regional fluxes with locally measured fluxes.

Carbon Dioxide

The observations were made in an investigation of the CO_2 flux over a wheat crop at Wagga Wagga, NSW on 15 d in the spring of 1991 (Raupach et al., 1992). At that time, the area around the experimental site supported winter cereals, peas (*Pisum sativum* L.), oilseed crops, and a mixture of improved and unimproved pastures. We expect the CO_2 assimilation by the wheat crop to have been broadly representative of the surrounding countryside. The free atmosphere CO_2 concentration was 351 ppm$_v$.

Examples of the local fluxes over the wheat crop and the CO_2 drawdown at 1.5 m on 2 days during the observations are given in Fig. 2–7. These show features observed on most days: highest rates of CO_2 assimilation in the morning (Fig. 2–7a), accompanied by rapid CO_2 drawdown with some overshoot, presumably because CO_2 was withdrawn from the CBL faster than it was replenished by entrainment, followed by a more gradual drawdown in the afternoon when the CBL grew slowly, if at all (Fig. 2–7b). In applying the integral budget method, Eq. [15], we chose $t = 0$ as the time when $C = C_+$, usually about 0800, and set the integration time as the period between $t = 0$ and 1500, that being the usual time of greatest CO_2 drawdown (Fig. 2-7b). The CBL height was set at 1 km, a typical spring value for Wagga Wagga deduced from historical radiosonde observations. The CBL CO_2 concentrations were estimated from the observations at 1.5 m by extrapolating to 20 m with the aid of an aerodynamic resistance model. Details of that calculation are given in Raupach et al. (1992).

Results of the analysis are shown in Fig. 2-8 that compares the regional integrated CO_2 fluxes with the locally measured fluxes. The fluxes from the CBL budget method are mostly smaller than those measured over the wheat crop (by 30% on average), but we expect this because the crop was dense, and high yielding. Importantly, the same day-to-day trends were evident in measured and modeled assimilation rates. Given the simplifying assumptions made in the analysis, the agreement is very encouraging.

Further developments in CBL budget methods are discussed by Raupach et al. (1992). They note that in reality, the integral approach described here only applies when the column of air in which the concentration measurements are made is within one day's travel from the sea where it will have had very nearly the same, free-atmosphere concentration throughout. Details of the trajectories of the air columns examined here are not presently available, but we note that Wagga Wagga is 300 to 500 km from the sea in the directions of the prevailing winds. It seems likely that on some days in the study, the air would have traveled more than 1 d from the sea, in which case the integral budget method would overestimate the regional flux. This would contribute to the scatter in Fig. 2-8.

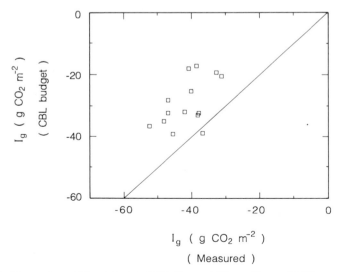

Fig. 2-8. Comparison of time-integrated CO_2 fluxes at Wagga Wagga, NSW, as measured over a wheat crop and as calculated by the integral CBL budget method, Eq. [15].

Nitrous Oxide

The observations were made at Canberra in 1978 in a period of 6 d after widespread rain. The local flux measurements, Fig. 2–9a, were made with
open chamber systems on grassland (Denmead et al., 1979). The corresponding atmospheric N_2O concentrations, Fig. 2–9b, were measured at 1.5 m
above the grass surface by continuous infrared gas analysis (Denmead, 1979b).
The surrounding countryside was predominantly grazed pasture. The value
of C_+ was taken to be 305 ppb$_v$. The points in Fig. 2–9b are hourly aver-

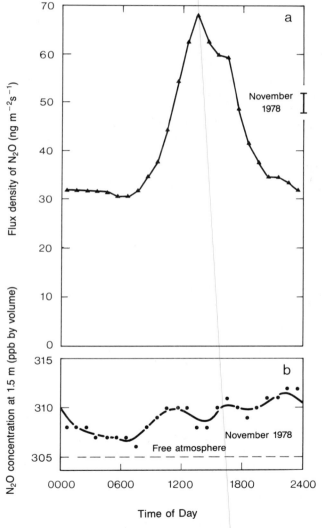

Fig. 2–9. (a) N_2O fluxes from grassland at Canberra during a 6-d period after rain (from Denmead et al., 1979); (b) corresponding averaged hourly N_2O concentrations at 1.5 m (from
Denmead, 1979b). Heavy lines are 3-h running means.

aged N_2O concentrations for the 6 d. The line is the 3-h running mean. Unlike CO_2, the N_2O flux was always positive so that N_2O concentrations reached minimum values with the start of convective mixing in the CBL shortly after sunrise and maximum values in the early evening. Like CO_2, however, the concentration change overshoots in the morning and then proceeds more gradually.

In this case, the integral budget approach is not applicable since we are dealing with data averaged over six consecutive days and we expect some carry-over in N_2O enrichment of the CBL from day to day. Instead, we have used the differential form of the CBL budget, Eq. [14]. Because we lack information on the rate of development of the CBL, we have restricted our analysis to the period between 0700 and 1030 when dC/dt was constant and we could expect the CBL to be growing rapidly. From developments in Raupach et al. (1992), we made an approximate analysis of the growth of the CBL during this period, based on energy input. From that analysis, we set the average depth of the CBL at 200 m and its growth rate at 100 m h^{-1}. Combining these values for L and dL/dt with the concentration changes shown in Fig. 9b, corrected for the probable difference between 1.5 m and the well-mixed CBL, we estimate the average regional flux from 0700 to 1030 to be 140 ng m^{-2} s^{-1}. The chamber measurements in Fig. 2–9a gave an average local N_2O flux of 37 ng m^{-2} s^{-1}, 26% of the regional estimate. Allowing for the crudeness of our assumptions and the large spatial variability of soil fluxes, the fact that the regional estimates and the spot measurements are of the same order of magnitude is heartening.

It is evident that CBL budget methods hold much promise. While certainly not definitive tests, the simple examples we have given illustrate the potential of this approach. Better measurement of C and accurate determination of L are needed to apply them successfully. Simultaneous radiosonde observations would obviously improve knowledge of L. In its simplest form, as in our examples, the general CBL approach appears to offer a very useful survey tool, giving order of magnitude estimates. The indications from Raupach et al. (1992) are that at best, the method could be a practical way to obtain detailed hour-by-hour information on regional trace gas fluxes over long time periods. Validation with aircraft data is highly desirable.

SUMMARY

The paper reviews methods for measuring the exchanges of trace gases between soil–plant systems and the atmosphere, with emphasis on the greenhouse gases N_2O, CH_4, and CO_2. The methods vary in scale from chambers covering < 1 m^2 to CBL budgets for regions of 100 to 1000 km^2. They should be seen as complementary rather than alternatives, each having a special role. Chambers permit replication and experiments with many land treatements, but can suffer from their interference to the microclimate and the large spatial variability of soil gas fluxes. Micrometeorological techniques are nondisturbing and integrate over larger areas. Mass balance methods can be used with small plots, up to 1 ha, for example. They are best suited to investigations where the background concentration of the gas is small and fluxes are large compared with normal emissions. Gradient and eddy correlation techniques integrate over areas of several hectares and can have high precision. Their application is limited presently by the availability of sensors that combine sensitivity, fast response, continuous operation, ruggedness, and low cost, and for eddy correlation, the need to apply large corrections to the apparent eddy flux. Two new micrometeorological approaches are filling gaps in our measurement capabilities: Lagrangian methods that offer a relatively simple means for inferring source–sink distributions of trace gases in plant canopies, and CBL budget methods that permit estimates of trace gas fluxes on regional scales, again with relatively simple measurements.

ACKNOWLEDGMENTS

We thank F.X. Dunin, R. Leuning, and W. Reyenga for access to their unpublished data and we acknowledge the material help provided by Charles Sturt University, the Grains Research and Development Corporation, and the Department of the Arts, Sport, Environment, Tourism and Territories in the conduct of the Wagga Wagga experiments.

REFERENCES

Baldocchi, D.D., B.B. Hicks, and T.P. Myers. 1988. Measuring biosphere–atmosphere exchanges of biologically related gases with micrometeorological methods. Ecology 69:1331–1340.

Beauchamp, E.G., G.E. Kidd, and G. Thurtell. 1978. Ammonia volatilization from sewage sludge applied in the field. J. Environ. Qual. 7:141–146.

Blackmer, A.M., S.G. Robbins and J.H. Bremner. 1982. Diurnal variability in rate of emission of nitrous oxide from soils. Soil Sci. Soc. Am. J. 46:937–942.

Businger, J.A., and S. Oncley. 1990. Flux measurement with conditional sampling. J. Atmos. Oceanic. Technol. 7:349–352.

Conrad, R. 1989. Control of methane production in terrestrial ecosystems. p. 39–58. In M.O. Andreae and D.S. Schimel (ed.) Exchange of trace gases between terrestrial ecosystems and the atmosphere. John Wiley & Sons, Chichester, England.

Denmead, O.T. 1979a. Chamber systems for measuring nitrous oxide emission from soils in the field. Soil Sci. Soc. Am. J. 43:89–95.

Denmead, O.T. 1979b. Surface measurements of atmospheric N_2O in S.E. Australia. p. 77. In Int. Assoc. Meteorol. Atmos. Physics, Program and Abstracts, Int. Union Geodesy Geophys., 7th Gen. Assembly, Canberra, Australia. 2–15 December. Natl. Center for Atmos. Res., Boulder.

Denmead, O.T. 1983. Micrometeorological methods for measuring gaseous losses of nitrogen in the field. p. 133–157. In J.R. Freney and J.R. Simpson (ed.) Gaseous loss of nitrogen from plant-soil systems. Martinus Nijhoff/Dr. W. Junk, The Hague.

Denmead, O.T. 1991. Sources and sinks of greenhouse gases in the soil-plant environment. Vegetatio 91:73–86.

Denmead, O.T., and E.F. Bradley. 1989. Eddy-correlation measurement of the CO_2 flux in plant canopies. p. 183–192. In Proc. 4th Australasian Conf. on Heat and Mass Transfer, Christchurch, New Zealand. 9–12 May. Secretariat, Fourth Australasian Conf. on Heat and Mass Transfer, Christchurch.

Denmead, O.T., J.R. Freney, and J.R. Simpson. 1979. Studies of nitrous oxide emission from a grass sward. Soil Sci. Soc. Am. J. 43:726–728.

Denmead, O.T., J.R. Simpson, and J.R. Freney. 1977. A direct field measurement of ammonia emission after injection of anhydrous ammonia. Soil Sci. Soc. Am. J. 41:827–828.

Desjardins, R.L., and J.I. MacPherson. 1989. Aircraft-based measurements of trace gas fluxes. p. 135–152. In M.O. Andreae and D.S. Schimel (ed.) Exchange of trace gases between terrestrial ecosystems and the atmosphere. John Wiley & Sons, Chichester, England.

Desjardins, R., P. Rochette, I. MacPherson, and E. Pattey. 1993. Measurements of greenhouse gas fluxes using aircraft- and tower-based techniques. p. 45–62. In L.A. Harper et al. (ed.) Agricultural ecosystem effects on trace gases and global climate change. ASA Spec. Publ. 55. ASA, CSSA, and SSSA, Madison, WI.

Dyer, A.J. 1963. The adjustment of profiles and eddy fluxes. Q.J.R. Meteorol. Soc. 89:276–280.

Folorunso, O.A., and D.E. Rolston. 1984. Spatial variability of field measured denitrification gas fluxes. Soil Sci. Soc. Am. J. 48:1214–1219.

Fowler, D., and J.H. Duyzer. 1989. Micrometeorological techniques for the measurement of trace gas exchange. p. 189–207. In M.O. Andreae and D.S. Schimel (ed.) Exchange of trace gases between terrestrial ecosystems and the atmosphere. John Wiley & Sons, Chichester, England.

Freney, J.R., O.T. Denmead, A.W. Wood, P.G. Saffigna, L.S. Chapman, G.J. Ham, A.P. Hurney, and R.L. Stewart. 1992. Factors controlling ammonia loss from trash covered sugarcane fields fertilised with urea. Fert. Res. 31:341–353.

Galbally, I.E., C.M. Elsworth, and H.A.H. Rabich. 1985. The measurement of nitrogen oxide (NO, NO_2) exchange over plant/soil surfaces. CSIRO Div. Atmos. Res. Tech. Pap. 8:1–23.

Gordon, R., M.Y. Leclerc, P. Schuepp, and R. Brunke. 1988. Field estimates of ammonia volatilisation from swine manure by a simple micrometeorological technique. Can. J. Soil Sci. 68:369–380.

Hicks, B.B., and R.T. McMillen. 1984. A simulation of the eddy accumulation method for measuring pollutant fluxes. J. Clim. Appl. Meteorol. 23:637–643.

Hutchinson, G.L. and G.P. Livingston. 1993. Use of chamber systems to measure trace gas fluxes. p. 63–78. In L.A. Harper et al. (ed.) Agricultural ecosystem effects on trace gases and global climate change. ASA Spec. Publ. 55. ASA, CSSA, and SSSA, Madison.

Hutchinson, G.L., and Mosier, A.R. 1979. Nitrous oxide emissions from an irrigated cornfield. Science (Washington, DC) 205:1225–1226.

Inoue, E., M. Tani, K. Imai, and S. Isobe. 1958. The aeorodynamic measurement of photosynthesis over the wheat field. J. Agric. Meteorol. (Japan) 13:121–125.

Jarvis, S. 1990. Nutrient flows and transfers. p. 48–56. In AFRC Institute of Grassland and Environmental Research Report for 1990. Hurley, U.K.

Jury, W.A., J. Letey, and T. Collins. 1982. Analysis of chamber methods used for measuring nitrous oxide production in the field. Soil Sci. Soc. Am. J. 46:250–256.

Keller, M., W.A. Kaplan, and S.C. Wofsy. 1986. Emissions of N_2O, CH_4 and CO_2 from tropical forest soils. J. Geophys. Res. 91:11 791–11 802.

Lemon, E.R. 1960. Photosynthesis under field conditions. II. An aerodynamic method for determining the turbulent carbon dioxide exchange between the atmosphere and a corn field. Agron. J. 52:697–703.

Lemon, E. 1978. Critique of "Soil and other sources of nitrous oxide": Nitrous oxide (N_2O) exchange at the land surface. p. 493–521. In D.R. Nielsen and J.G. MacDonald (ed.) Nitrogen in the environment. Vol. 1. Nitrogen behavior in field soil. Academic Press, New York.

Lenschow, D.H., and M.R. Raupach (1991). The attenuation of fluctuations in scalar concentrations through sampling tubes. J. Geophys. Res. 96:15 259–15 268.

Leuning, R., J.R. Freney, O.T. Denmead, and J.R. Simpson. 1985. A sampler for measuring atmospheric ammonia flux. Atmos. Environ. 19:1117–1124.

Leuning, R., and K.M. King. 1992. Comparison of eddy-covariance measurements of CO_2 fluxes by open- and closed-path CO_2 analysers. Boundary-Layer Meteorol. 59:297–311.

Leuning, R., and J. Moncrieff. 1990. Eddy co-variance CO_2 flux measurements using open- and closed-path analysers: Corrections for analyser water vapour sensitivity and damping of fluctuations in air sampling tubes. Boundary-Layer Meteorol. 53:63–76.

Matthias, A.D., A.M. Blackmer, and J.M. Bremner. 1980. A simple chamber technique for field measurement of emissions of nitrous oxide from soils. J. Environ. Qual. 9:251–256.

Monteith, J.L., and G. Szeicz. 1960. The carbon dioxide flux over a field fo sugar beet. Q.J.R. Meteorol. Soc. 86:205–214.

Monteith, J.L., G. Szeicz, and K. Yabuki. 1964. Crop photosynthesis and the flux of carbon dioxide below the canopy. J. Appl. Ecol. 1:321–337.

Mosier, A.R. 1989. Chamber and isotope techniques. p. 175–187. *In* M.O. Andreae and D.S. Schimel (ed.) Exchange of trace gases between terrestrial ecosystems and the atmosphere. John Wiley & Sons, Chichester, England.

Mulhearn, P.J. 1977. Relations between surface fluxes and mean profiles of velocity, temperature and concentration, downwind of a change in surface roughness. Q.J.R. Meteorol. Soc. 103:785–802.

Nouchi, I., S. Mariko, and K. Aoki. 1990. Mechanism of methane transport from the rhizosphere to the atmosphere through rice plants. Plant Physiol. 94:59–66.

Philip, J.R. 1959. The theory of local advection: 1. J. Meteorol. 16:535–547.

Raupach, M.R. 1989a. A practical Langrangian method for relating scalar concentrations to source distributions in vegetation canopies. Q.J.R. Meteorol. Soc. 115:609–632.

Raupach, M.R. 1989b. Applying Langrangian fluid mechanics to infer scalar source distributions from concentration profiles in plant canopies. Agric. For. Meteorol. 47:85–108.

Raupach, M.R. 1991. Vegetation-atmosphere interaction in homogeneous and heterogeneous terrain: Some implications of mixed-layer dynamics. Vegetatio 91:105–120.

Raupach, M.R., O.T. Denmead, and F.X. Dunin. 1992. Challenges in linking surface CO_2 concentrations to fluxes at local and regional scales. Aust. J. Bot. 40:697–716.

Raupach, M.R., and B.J. Legg. 1984. The uses and limitations of flux-gradient relationships in micrometeorology. Agric. Water Manage. 8:119–131.

Rennenberg, H. 1992. Trace gas exchange in rice cultivation. Ecol. Bull. 42:(In press.)

Schütz, H., and W. Seiler. 1989. Methane flux measurements: Methods and results. p. 209–228. *In* M.O. Andreae and D.S. Schimel (ed.) Exchange of trace gases between terrestrial ecosystems and the atmosphere. John Wiley & Sons, Chichester, England.

Seiler, W., A. Holzapfel-Pschorn, R. Conrad, and D. Scharffe. 1984. Methane emission from rice paddies. J. Atmos. Chem. 1:241–268.

Sutton, O.G. 1953. Micrometeorology. McGraw Hill, New York.

Wahlen, M., N. Tanaka, R. Henry, B. Deck, J. Seglen, J.S. Vogel, J. Southon, A. Shemesh, R. Fairbanks, and W. Broecker. 1989. Carbon-14 in methane sources and in atmospheric methane: The contribution from fossil carbon. Science (Washington, DC) 245:286–290.

Watson, R.T., H. Rohde, H. Oeschger, and U. Siegenthaler. 1990. Greenhouse gases and aerosols. p. 1–40. J.T. Houghton et al. (ed.) Climate change: The IPCC scientific assessment. Intergovernmental Panel on Climate Change, Cambridge Univ. Press, Cambridge, England.

Webb, E.K., G.I. Pearman, and R. Leuning. 1980. Correction of flux measurements for density effects due to heat and water vapour transfer. Q.J.R. Meteorol. Soc. 106:85–100.

Wesely, M.L., J.A. Eastman, D.H. Stedman, and E.D. Yalvac. 1982. An eddy-correlation measurement of NO_2 flux to vegetation and comparison to O_3 flux. Atmos. Environ. 16:815–820.

Wesely, M.L., D.H. Lenschow, and O.T. Denmead. 1989. Flux measurement techniques. p. 31–46. *In* D.H. Lenschow and B.B. Hicks (ed.) Global tropospheric chemistry: Chemical fluxes in the global atmosphere. NCAR, Boulder, CO.

Wilson, J.D., V.R. Catchpoole, O.T. Denmead, and G.W. Thurtell. 1983. Verification of a simple micrometeorological method for estimating the rate of gaseous mass transfer from the ground to the atmosphere. Agric. Meteorol. 29:183–189.

Wilson, J.D., G.W. Thurtell, G.E. Kidd, and E.G. Beauchamp. 1982. Estimation of the rate of gaseous mass transfer from a surface plot to the atmosphere. Atmos. Environ. 16:1861–1867.

Wilson, J.D., and W.K.N. Shum. 1992. A re-examination of the integrated horizontal flux method for estimating volatilisation from circular plots. Agric. For. Meteorol. 57:281–295.

3 Measurements of Greenhouse Gas Fluxes Using Aircraft- and Tower-based Techniques

Raymond Desjardins, Philippe Rochette, and Elizabeth Pattey

Agriculture Canada
Centre for Land and Biological Resources Research
Ottawa, Canada

Ian MacPherson

National Research Council of Canada
Ottawa, Canada

The atmospheric concentration of radiatively active gases, such as CO_2, CH_4, N_2O, and O_3, is increasing at an alarming rate. This increase is expected to alter the Earth's climate, which could have serious consequences on the agricultural sector (Taylor & MacCracken, 1990). Agricultural activities are not the main source of the observed increase of greenhouse gases. However, since agroecosystems are the most intensively managed nonurban part of the globe, they are very amenable to management decisions regarding greenhouse gas emissions and absorption. They could, therefore, play an important role in our efforts to limit the increase in concentration of these gases (Bouwman, 1990).

Sources of greenhouse gases in the agricultural sector are:

1. Carbon dioxide through the use of fossil fuels, loss of soil organic matter and biomass burning.
2. Methane from ruminants, rice (*Oryza sativa* L.) production, biomass burning, agricultural wastes and wet soils.
3. Nitrous oxide from fossil fuel and biomass burning, fertilizer use and intensive cultivation.
4. Ozone through the production of volatile organic compounds, CH_4, CO and NO_x.

Several approaches appear promising to increase the sinks or decrease the sources of greenhouse gases: (i) energy conservation techniques, (ii) increasing the amount of soil organic matter through innovative cropping systems, (iii) reduction of CH_4 output from domestic livestock and rice production, and (iv) reduction of N_2O emissions associated with high fertilizer applications. It is very important to consider the interactions among trace gases because of their different efficiency in trapping infrared radiation. For example, the value of management practices that increase organic matter in the soil could be lost if by doing so the N_2O emissions to the atmosphere were increased (Bohn, 1990).

Accurate quantification of the sources and sinks of greenhouse gases from agroecosystems are needed to orient research efforts towards the right targets and to evaluate the impact of the proposed solutions. The sources and sinks of greenhouse gases are reasonably well identified but their magnitudes are not well known (Bouwman, 1990). This is due to the complexity of biological systems, the difficulty of measuring gas exchange under field conditions, and the large spatial and temporal variability of greenhouse gas emissions (Duxbury et al., 1982). Several techniques have been used to characterize the sources and sinks of greenhouse gases (Stewart et al., 1989). Chambers and micrometeorological techniques are most commonly used and are well reviewed in the literature (Businger, 1986; Baldocchi et al., 1988; Fowler & Duyzer, 1990; Mosier, 1989). This essay will not duplicate these efforts nor will it discuss the use of gradient techniques that will be reviewed by Denmead and Raupach (1992). It will examine techniques that can be used both on tower- and aircraft-based systems. Most of the discussion will concentrate on the eddy-correlation technique, but alternative methods such as the eddy-accumulation, variance and dissipation techniques also will be briefly examined.

MEASURING TECHNIQUES

Eddy Correlation

The eddy-correlation technique provides the most direct measurement of the flux of a gas at the land–atmosphere interface. It provides measurements of gas exchange without disturbing the environment under study and integrates the flux over a mosaic of different sources and sinks. The instantaneous transfer across a horizontal plane per unit area and per unit time is given by

$$F = WS$$

where W is vertical velocity, and S is the mixing ratio of the gas of interest. Because turbulent transfer is intermittent, it must be averaged over a certain time or distance to obtain a representative sample. The mean flux of a gas over a horizontally homogeneous surface under steady-state conditions is given by

$$F = \overline{W'S'}$$

where W' and S' are the fluctuations from their mean. Reasonably stationary conditions, little horizontal advection and no chemical reaction involving the gas of interest within the air column below the measuring system are required for accurate flux measurements. These effects need to be taken into account to minimize flux divergence with height, which can lead to significant errors of surface flux estimates from aircraft-based systems (Betts et al., 1990).

To date, the eddy-correlation technique has been used to measure the flux of the following greenhouse gases: CO_2, H_2O, O_3, and CH_4 (Lenschow et al., 1982; Desjardins et al., 1989; Chahuneau et al., 1989; Fan et al., 1982; Gusten et al., 1992); and, of gases involved in their production and/or consumption—NO, NO_2, and CO (Wesely et al., 1982; Delany et al., 1986). Nitrous oxide measurements have not been obtained yet with this technique because of the lack of fast-response sensors with adequate sensitivity.

Eddy Accumulation

The eddy-accumulation technique is an alternative measuring technique not requiring fast-response chemical sensors (Desjardins, 1972; Desjardins et al., 1984). It involves collecting air proportionally to the vertical wind, and obtaining the mean concentration of the gas of interest of the upward and downward moving air over the sampling period. This approach avoids the need of fast-responding chemical sensors but requires a fast-response vertical wind sensor, proportional sampling valves (Buckley et al., 1988), and very sensitive analyzers for detecting small differences in gas concentration. This is not always possible because of the detection limits of the sensors available (Table 3–1). For example, Hicks and McMillen (1984) have estimated that in order to achieve a deposition velocity (Fc/ρ_c) of ± 1 mm s^{-1}, a relative accuracy of the chemical sensors of $\pm 0.4\%$ was required.

More recently, a so-called relaxed eddy-accumulation technique has been proposed by Businger (1986), Businger and Oncley (1990), and adapted to aircraft-based systems by MacPherson and Desjardins (1991). In this technique, a valve splits a constant sampling flow rate strictly on the basis of

Table 3–1. Information on sensor characteristics for CO_2, O_3, CH_4, and N_2O.

Species	Technique	Time constant	Detection limit
CO_2	IR absorption	0.05 s	0.20 ppmv
	IR absorption	2 s	0.03 ppmv
O_3	NO chemiluminescence	0.05 s	0.5 ppbv
	Ethylene chemiluminescence	0.5 s	0.5 ppbv
	UV absorption	2 s	2.0 ppbv
CH_4	Tunable diode laser	0.1 s to min	5 ppbv
	IR absorption	0.1 s to min	10 ppbv
	GC with FID†	1 s to min	10 ppbv
N_2O	GC with ECD‡	min	2 ppbv

† FID = flame ionization detection.
‡ ECD = electron capture detection.

the direction of vertical wind. The flux is then given by the following relationship

$$F = A \, \sigma_W \, (\overline{C^+} - \overline{C^-})$$

where A is an empirical constant, which is approximately 0.6; σ_W is the standard deviation of the vertical wind; and $(\overline{C^+} - \overline{C^-})$ is the difference between the mean concentrations of the upward and downward moving air. This approach appears to be the only technique that could be used on an aircraft to measure the flux of gases for which there are no fast-responding sensors such as volatile organic components, etc.

Variance and Dissipation

Two other techniques requiring only fast-response chemical sensors are sometimes used to estimate fluxes. These are the variance and the dissipation techniques.

The variance technique, which has been extensively described by Wesely (1988), assumes that the correlation coefficients r for vertical wind (W) and any scalar are identical; that is, $r_{WT} = r_{Wq} = r_{WC}$ where T is air temperature, q is the water vapor concentration, and C is the concentration of any gas of interest. By definition the turbulent flux of C is expressed

$$F_C = r_{WC} \, \sigma_W \, \sigma_C$$

$$\text{since } r_{Wq} = F_q/(\sigma_W \, \sigma_q)$$

$$\text{then } F_C = (F_q/\sigma_q) \, \sigma_C$$

If estimates of F_q are unavailable, a value of the correlation coefficient r_{Wq} must be assumed. This approach provides no information on the direction of the flux. Nowadays it is not often used since vertical wind sensors are readily available.

The inertial-dissipation technique can also provide flux estimates without sophisticated vertical wind measurements. This is particularly useful on boats and small aircraft where it is difficult to measure vertical wind velocity accurately. Only the inertial subrange characteristics of the spectrum of the scalar of interest are required (Fairall & Larsen, 1986). These spectral characteristics are related to flux through the equations for the evolution of turbulent kinetic energy, and temperature and specific humidity concentration. These equations, given by Durand et al. (1991), are solved using an iterative process. Reasonable results have been reported by several groups for momentum, sensible and latent heat fluxes from towers and from low-level aircraft measurements.

SENSORS

Vertical Wind Sensors

Several types of sonic anemometers are now commercially available to measure fast-response fluctuations of air motion on towers. They provide excellent measurements of turbulence as long as the sensors are at least a few meters above the surface of interest and factors such as shadow effect, flow distortion by obstacles, and coordinate transformation are taken into account (Wyngaard, 1988). On aircraft, air turbulence has been measured using several sensors such as differential pressure probes on a boom or on the nose, fixed vane and gust probe with light pivoting vanes (Desjardins & MacPherson, 1989; Lenschow et al., 1991). Errors resulting from the inadequate compensation for aircraft motion effect during maneuvering flight have been estimated to typically add about 0.1 m s^{-1} to actual standard deviations in vertical velocity. This corresponds to an error of ± 10 to 20% in the flux estimates. During straight level runs, typical errors are a factor of five smaller and turbulence measurements are not considered to be the limiting factor in measuring turbulent fluxes (Desjardins & MacPherson, 1991).

Chemical Sensors

For a long time, the lack of sensitive, reliable, and fast-responding sensors has been the main limiting factor in using the eddy-correlation technique for measuring greenhouse gas fluxes. Several sensors are now available (Table 3–1) for both tower and aircraft flux measurements of O_3, CO_2, and CH_4, (Lenschow & Hicks, 1989). As mentioned above, no fast-response sensor is commercially available for N_2O, but the development of tunable diode laser technology should soon make available the measurement of flux of this gas (Sachse et al., 1987; G.W. Thurtell, 1991, personal communication). Sensitivity requirements of sensors for measuring typical fluxes of these gases can be estimated from the mean difference in concentration between updrafts and downdrafts for typical flux values (Table 3–2). A resolution of one-tenth of this difference is required in order to make satisfactory flux measurements. This is readily obtained by CO_2 and O_3 sensors, but existing CH_4 sensors need to be improved.

Table 3–2. Mean difference in concentration between updrafts and downdrafts for typical flux values of four greenhouse gases.

Species	Typical flux values		Mean difference	
CO_2	-2	mg m^{-2} s^{-1}	4	ppmv
CH_4	1	μg m^{-2} s^{-1}	5	ppbv
N_2O	0.4	μg m^{-2} s^{-1}	0.7	ppbv
O_3	-1	μg m^{-2} s^{-1}	2	ppbv

FACTORS AFFECTING FLUX MEASUREMENT ACCURACY

This section will primarily concentrate on the eddy-correlation technique because it is the most direct method of estimating fluxes. Many factors can affect the accuracy of eddy-flux measurements. The loss of measurement accuracy of vertical wind associated with the motion on boats and planes is a problem that is difficult to assess, but where considerable progress has been achieved in recent years (Crawford et al., 1991; Lenschow et al., 1991). Errors due to flow distortions around the turbulence sensors by tower, aircraft or other sensors must be minimized as much as possible because they cannot be corrected for (Wyngaard, 1988). Measurements over nonlevel terrain result in biased-mean vertical wind velocity. This effect can usually be dealt with coordinate rotation if the slope of the terrain is less than 15% (Baldocchi et al., 1988). Nonstationary conditions and horizontal advection, which can contribute to flux divergence with height, must be considered in extrapolating aircraft flux measurements to the surface (Betts et al., 1990). Other factors, such as inadequate sensor response characteristics (Moore, 1986), lag effects due to sensor separation, density corrections due to sensible and latent heat fluxes and inadequate fetch characteristics will be briefly discussed.

Cospectral Characteristics

Cospectral analysis is a useful way to characterize the contribution of various frequencies or wavenumbers to the flux. It provides information on whether the sensors response time is adequate or if the sampling height is too low to characterize most of the gas transfer. Cospectra and the cumulative cospectra of vertical wind, W, with CO_2 (C_{WC}) and vertical wind with H_2O (C_{Wq}) as a function of wavenumber for typical aircraft-based measurements using fast- and slow-response CO_2 and H_2O analyzers are presented in Fig. 3–1 and 3–2. Each cospectrum is based on the average of three 35-km runs. The fast-response analyzer for CO_2 and H_2O (ESRI, open-path infrared analyzer built by Agriculture Canada, Ottawa, ON) had a frequency response of 7 Hz and the slow-response CO_2 (LI-COR 6250, LI-COR, Lincoln, NE) and H_2O (Dew point hygrometer, EGG, Environmental Equipment Div., Waltham, MA) analyzers had a frequency response of 0.5 and 0.2 Hz, respectively. The cumulative plots of the cospectra show that in this case the slow-response sensors underestimate the fluxes by approximately 20% at an altitude of 30 m while, at 200 m, the response of these sensors is adequate to measure most of the flux contribution. The contribution of long wavelengths to the flux can be underestimated by an aircraft-based system because of too short or too high runs. Cospectral estimates in Fig. 3–1 and 3–2 indicate that, at 200 m, 50% of the flux comes from wavenumbers smaller than 10^{-3} m^{-1}, which correspond to wavelengths longer than 1 km.

The importance of long wavelengths can also be determined by removing long wavelength contributions using high-pass filtering. When high-pass filtered (5-km wavelength) and linearly detrended O_3 and CO_2 fluxes were recorded for 47 15-km runs collected on four different days, at an altitude of approximately 30 m, no significant contribution from wavelengths longer than the cutoff frequency of 5 km was observed (Fig. 3–3a,b). In other cases such as over a grasslands study with changes in elevations of up to 80 m, detrended fluxes were on the average 25% larger than filtered (high-pass with a 5-km cutoff) fluxes. Such long wavelength contributions demonstrate that under certain circumstances relatively long runs are required in order to estimate surface fluxes from aircraft-based systems.

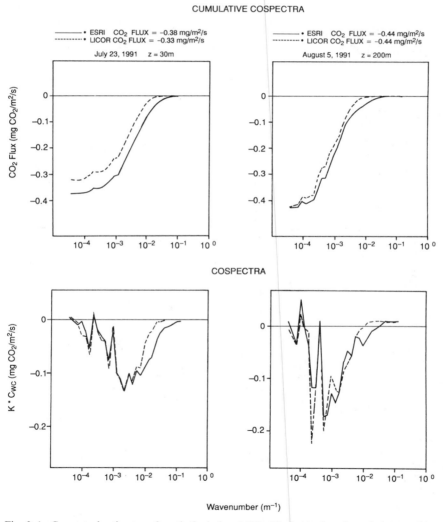

Fig. 3–1. Cospectral estimates of vertical wind and CO_2 (C_{wc}) as a function of wavenumber for measurements recorded at 30 and 200 m with two different CO_2 analyzers.

Lag Effect

Flow distortion around vertical wind sensors can cause considerable errors in flux measurements. To avoid this problem, sensors are often separated to measure the same eddies with a time delay. The effect of this time delay can be counteracted by shifting one time series with respect to the other in the analysis procedure. It is more difficult to correct for lag effects with open-path gas analyzers on towers because of fluctuating wind speed and direction.

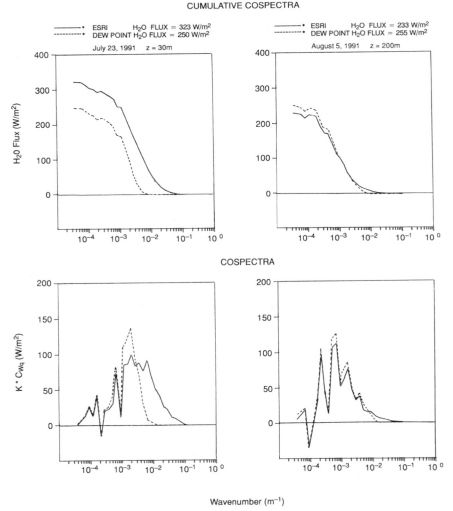

Fig. 3-2. Cospectral estimates of vertical wind and water vapor (C_{wq}) as a function of wavenumber for measurements recorded at 30 and 200 m with two different water vapor analyzers.

Time delays due to ducting the air sample or due to separation of the species and vertical velocity sensors can also cause significant underestimations of fluxes on aircraft and tower due to loss of correlation between W and the scalar of interest. The appropriate time delay can be determined by examining the variations in the correlation coefficients for various lags: W and CO_2 and, air temperature are correlated for delays of 0 to 10 data samples, where one data point corresponds to 1/16 (0.0625) of a second (Fig. 3–4). These data were collected at four altitudes using an aircraft-based system and the sensors were mounted at three different locations on the aircraft. The degree of underestimation is proportional to the change in the correlation coefficients. The lag that gave maximum correlation corresponds very well to the time for the air to reach the various sensors. For example,

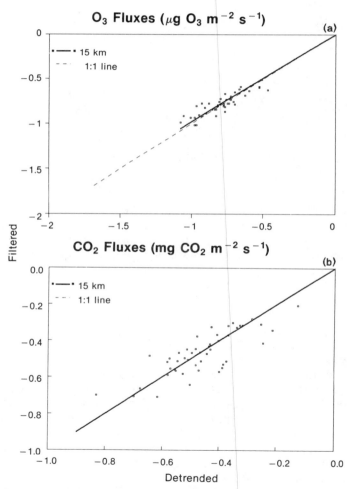

Fig. 3–3. Relationship between filtered vs. linearly detrended fluxes of (a) O_3 and (b) CO_2 over agricultural regions in California.

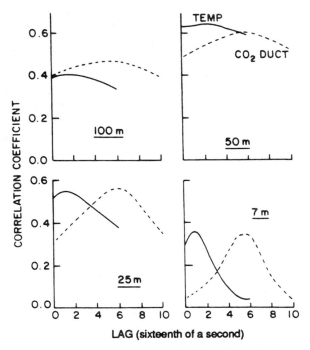

Fig. 3-4. Correlation coefficients between vertical wind and lagged air temperature (TEMP) and, vertical wind and lagged CO_2 (CO_2 duct) for four elevations.

the temperature sensor that was the closest to the vertical wind sensor has a maximum correlation at a lag of 1/16 (0.0625) s (Fig. 3-4). As expected, the lower the sampling height, the more critical this effect becomes because the eddies contributing to the transfer are smaller closer to the surface. A time delay of 3/16 (0.1875) s can cause an underestimation of 50 and 10% at altitudes of 7 and 100 m, respectively.

Air Density Fluctuations

Another problem is the correction of the trace gas density due to water vapor and heat transfer. For several gases such as CO_2, this correction can not only change the magnitude but also the sign of the flux. Webb et al. (1980) demonstrated how to use the latent and sensible heat fluxes to correct for density variations. As a general rule, micrometeorological techniques yield erroneous flux measurements during typical daytime conditions if the ratio of the mean concentration to flux is greater than 70 s m^{-1} (Pattey et al., 1992). Such corrections are not necessary if the concentration is measured in terms of mixing ratio, that is, mass of gas of interest over the mass of dry air. Some sensors measure mixing ratio directly; however, most sensors only measure gas density. Carbon dioxide flux measurement estimated using density corrections based on sensible and latent heat fluxes and using the

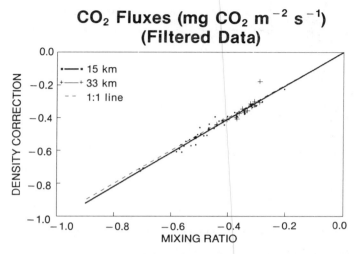

Fig. 3–5. Relationship between CO_2 fluxes based on Webb's density correction and CO_2 fluxes based on calculated mixing ratio with respect to dry air for runs of 15 and 33 km.

calculated mixing ratio of CO_2 are presented in Fig. 3–5. The mixing ratio was calculated based on the CO_2 density fluctuations and the pressure, temperature and moisture fluctuations in the open path of the analyzer. These results show an agreement within 5%. For gases such as N_2O, where the magnitude of the correction is as large as the flux itself, it will be important to find ways to minimize the need of this correction by drying the air and allowing it to reach thermal equilibrium rapidly.

Footprint Characteristics

The flux footprint is the area contributing to the flux observed at one point. This area has been estimated using several empirical models (Schuepp et al., 1990; Horst & Weil, 1992). It is dependent on the friction velocity, U_*, surface roughness, Z_o, the height of the vegetation, d, and the sampling height, z. The footprint area sampled is considerably smaller for unstable, than for stable conditions, and for rough surfaces compared with smooth surfaces. Because of the intermittency of turbulent transport and nonsteady wind conditions the footprint sensed by a flux measuring system is at best an ensemble average of instantaneous point source releases. For a $U_* = 0.4$ m s^{-1}, $Z_o = 0.06$ m, and $d = 0.3$ m; these models predict that when sampling at 5 m, 90% of the flux contribution comes from a distance within 1 km, while at 30 m only 25% of the flux contribution originates within the same distance (Fig. 3–6). Such results need further evaluation but they provide valuable information on the area sampled by flux measuring systems.

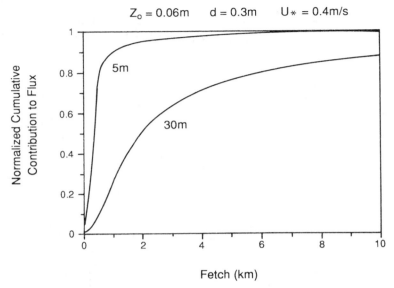

$Z_0 = 0.06m$ $d = 0.3m$ $U_* = 0.4m/s$

Fetch (km)

Fig. 3–6. Normalized cumulative contribution to flux at 5 and 30 m for various fetch for a specific roughness length Z_0, displacement height d, and a friction velocity, U_*.

MEASURING SYSTEMS

The main difficulty in quantifying the role of agroecosystems in the increase of greenhouse gases in the atmosphere is to obtain averages of flux estimates in space and time. Tower-based measurements provide diurnal averages over areas of a few thousand square meters while aircraft-based measurements provide spatial averages over areas a thousand times larger but for only a few hours at a time. For these reasons, it is important to use both systems if we are to characterize the dynamics of complex ecosystems.

Tower-based Measurements

Tower-based measurements provide diurnal patterns of fluxes of gases that are site specific. Measurements of the covariance between vertical wind and the CO_2 concentration a few meters above the surface have been obtained above agricultural crops using tower-based systems for the last 20 yr (Desjardins, 1992). Ozone and CH_4 fluxes have also been measured extensively with this technique by several research groups (Wesely et al., 1982; Ogram et al., 1988; Fan et al., 1992).

Flux of trace gases should be measured concurrently with fluxes of sensible and latent heat in order to correct for density variations if necessary, and to verify the accuracy of the flux measurements based on energy balance considerations. During relatively calm conditions, eddy-flux measurements are sometimes considerably underestimated. Figure 3–7 shows the net radia-

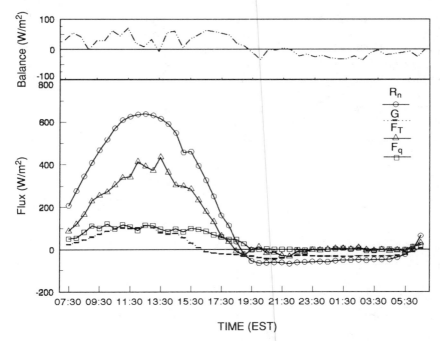

Fig. 3-7. Example of tower-based measurements of net radiation, R_n, soil heat flux, G, sensible, F_T, and latent heat, F_q, fluxes obtained at 3 m above a cereal crop in Ottawa. The balance is $R_n - G - F_T - F_q$.

tion, R_n, soil heat flux, G, corrected for heat storage in the top 50 mm of soil as well as the sensible, F_T and latent, F_q, heat fluxes measured at 3 m above a cereal crop in Ottawa, ON, Canada. The available energy at the soil surface $(R_n - G)$ minus the energy lost through sensible and latent heat fluxes indicates that, in this case, the fluxes of heat and water vapor were underestimated over the 24-h period by approximately 10%. It is not clear whether this is due to nonrepresentative measurements of R_n and G, which are very local measurements, or an underestimation of the turbulent fluxes. Many factors can lead to an underestimation of turbulent fluxes such as spatial averaging by sensors due to sensing path length, separation between W and chemical sensor and any other factors that cause a reduction in the correlation between the vertical wind and the scalar of interest.

Aircraft-based Measurements

Airborne-flux measurements provide a very promising way to estimate greenhouse gas exchange over agricultural regions. They provide aerial averages that are comparable in scale to satellite observations. Desjardins et al. (1992) have shown very high correlations between CO_2 fluxes and satellite spectral data for a 15- by 15-km grassland area.

Several aircraft flux systems have been developed in recent years (Lenschow, 1986). The Flight Research Laboratory of the National Research Council (NRC), Agriculture Canada and McGill University have cooperated for the last 10 yr on the development of instrumentation and analysis techniques to make airborne measurements of the fluxes of heat, momentum, water vapor, CO_2, O_3, and CH_4 (Desjardins & MacPherson, 1989, 1991). MacPherson et al. (1992) have extensively compared flux results from three aircraft during several wing-to-wing intercomparison flights. Figure 3–8 shows a typical example of the results obtained over a distance of 75 km (wing-to-

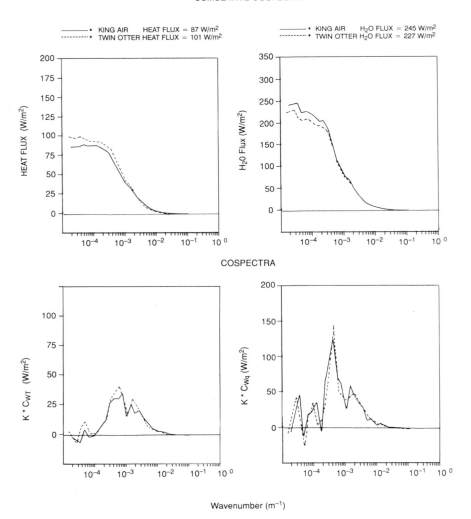

Fig. 3–8. Cospectral estimates of vertical wind and temperature, C_{WT}, and water vapor, C_{Wq}, obtained at 150 m over a distance of 75 km with two aircraft flying wing-to-wing.

wing) at an altitude of 150 m between Twin Otter and the King Air from the National Center for Atmospheric Research in Boulder, CO. The agreement is well within the expected range considering that the two aircraft are not sampling exactly the same air.

As discussed in a previous section, many factors can affect the accuracy of aircraft-based flux measurements. Flux divergence in the first 100 m can sometimes account for 10% error in the surface flux estimate (Betts et al., 1990). If the run length is short compared to the size of the transporting eddies considerable uncertainty also can result. A cross-wind flight path of 25 km, over relatively homogeneous terrain, is thought to be required for an error equal to or lower than 10% on fluxes measured by an aircraft flying at 150 m above ground (Marht & Gibson, 1992). Most of the sources of errors have been identified and it is now considered feasible to obtain accurate estimates of surface fluxes for several gases such as CO_2 and H_2O using aircraft-based sensors (Desjardins et al., 1992).

Figure 3–9 shows an example of flux measurements of CO_2, H_2O, and O_3 over mixed farmland in California at an altitude of 32 m. These values are 1-km averages with a 3-km running mean. The greenness index that is measured directly under the aircraft along a 4-m strip is also shown to give an indication of the amount of green vegetation overflown. This measurement is based on the reflected near infrared–red radiation ratio. A ratio of 1 normally indicates bare soil, while a value of 5 is typical of very lush vegetation. This ratio has been multiplied by 20 to put it in on a comparable scale to the other flux values. For example the evaporation and CO_2 uptake are

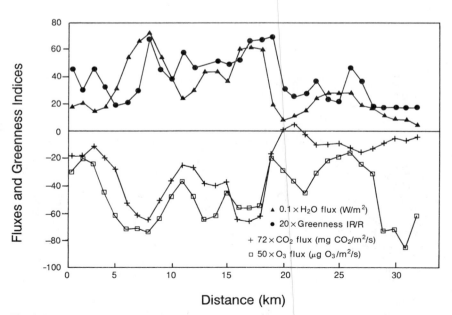

Fig. 3–9. Carbon dioxide, water, and ozone fluxes as well as greenness index all measured at 32 m above an agricultural region in the San Joaquin Valley on 30 July 1991.

large where the vegetation is very green. Larger O_3 uptake also tends to coincide with crop activity such as photosynthesis and transpiration but substantial O_3 fluxes are also observed over bare soil near the end of the run. These measurements are presented mainly as an example of the type of data now available to develop models for estimating trace gas exchanges. Such models are essential if we are to arrive at a regional budget of greenhouse gases over long time periods (Hicks et al., 1987).

CONCLUSIONS

Much progress has been achieved in recent years in using aircraft and tower-based systems to measure the flux of trace gases. These measurements have been useful in characterizing the dynamics of several ecosystems, such as grasslands and wetlands, but their ultimate value will be to provide data to develop models that can be used to extrapolate to conditions that have not, and cannot be studied. The combination of tower- and aircraft-based measurements appears promising for providing diurnal patterns of area average fluxes. Improvements in sensors are still required particularly for CH_4 and N_2O. For other gases such as volatile organic compounds, the most promising approach is the eddy-accumulation technique.

As far as agroecosystems are concerned, more concerted efforts using enclosure, tower- and aircraft-based systems are required to quantify the sources and sinks of greenhouse gases.

REFERENCES

Baldocchi, D.D., B.B. Hicks, and T.P. Meyers. 1988. Measuring biosphere–atmosphere exchanges of biologically related gases with micrometeorological methods. Ecology 69:1331–1340.

Betts, A.K., R.L. Desjardins, J.I. MacPherson, and R.D. Kelly. 1990. Boundary-layer heat and moisture budgets from FIFE. Boundary Layer Meteorol. 50:109–138.

Bohn, H.L. 1990. Considerations for modelling carbon interactions between soil and atmosphere soils and the greenhouse effect. p. 391–394. In A.E. Bouwman (ed.) Soils and the greenhouse effect. John Wiley & Sons, Ltd., Toronto, Canada.

Bouwman, A.E. 1990. Exchange of greenhouse gases between terrestrial ecosystems and the atmosphere. p. 61–129. In A.E. Bouwman (ed.) Soils and the greenhouse effect. John Wiley and Sons, Toronto.

Buckley, D.J., R.L. Desjardins, J.L.M. Lalonde, and R. Brunke. 1988. A linearized fast-response gas sampling apparatus for eddy accumulation studies. Comput. Electron. Agric. 2:243–250.

Businger, J.A. 1986. Evaluation of the accuracy with which dry deposition can be measured with current micrometeorological techniques. J. Clim. Appl. Meteorol. 25:1100–1124.

Businger, J.A., and S.P. Oncley. 1990. Flux measurements with conditional sampling. J. Atmos. Oceanic Technol. 7:349–352.

Chahuneau, F., R.L. Desjardins, E.J. Brach, and R. Verdon. 1989. A micrometeorological facility for eddy flux measurements of CO_2 and H_2O. J. Atmos. Oceanic Technol. 6:193–200.

Crawford, T.L., R.T. McMillen, and R.J. Dobosy. 1991. Description of a generic mobile flux platform using a small airplane and a pontoon boat. p. 37–41. In 7th Symp. Meteorological Observations and Instrumentation, New Orleans. 13–18 Jan. 1991. Am. Meteorol. Soc., Boston, MA.

Delany, A.C., D.R. Fitzjarrald, D.H. Lenschow, R. Pearson, Jr., G.J. Wendel, and B. Woodruff. 1986. Direct measurements of nitrogen oxides and ozone fluxes over grassland. J. Atmos. Chem. 4:429–444.

Desjardins, R.L. 1972. A study of carbon dioxide and sensible heat fluxes using the eddy correlation technique. Ph.D. diss. Cornell Univ., Ithaca, NY (Diss. Abstr. 73-34).

Desjardins, R.L. 1992. Review of techniques to measure CO_2 flux densities from surface and airborne sensors. p. 1–23. In G. Stanhill (ed.) Monogr. on advances in bioclimatology. Vol. 1. Springer-Verlag, Ltd.

Desjardins, R.L., D. Buckley, and G. St.-Amour. 1984. Eddy flux measurement of CO_2 using a microcomputer system. Agric. For. Meteorol. J. 32:257–265.

Desjardins, R.L., and J.I. MacPherson. 1989. Aircraft-based measurements of trace gas fluxes. p. 135–152. In M.O. Andrea and D.S. Schimel (ed.) Exchange of trace gases between terrestrial ecosystems and the biosphere. John Wiley and Sons, Ltd., Chester, England.

Desjardins, R.L., and J.I. MacPherson. 1991. Water vapor flux measurements from aircraft. p. 245–260. In T. Schmugge and J.C. André (ed.) Workshop Proc. on the Measurement and Parameterization of Land-Surface Evaporation Fluxes, Banyuls, France. October 1988. Spring-Verlag, Ltd., Heidelberg, Germany.

Desjardins, R.L., J.I. MacPherson, P.H. Schuepp, and F. Karanja. 1989. An evaluation of airborne eddy flux measurements of CO_2, water vapor, and sensible heat. Boundary Layer Meteorol. 47:55–69.

Desjardins, R.L., P.H. Schuepp, J.I. MacPherson, and D.J. Buckley. 1992. Spatial and temporal variations of the fluxes of carbon dioxide and sensible and latent heat over the FIFE site. J. Geophys. Res. 97:18 467–18 475.

Denmead, O.T., and M.R. Raupach. 1993. Methods for atmospheric measuring gas transport in agricultural and forest systems. p. 19–43. In L.A. Harper et al. (ed.) Agricultural ecosystem effects on trace gases and global climate change. ASA Spec. Publ. 55. ASA, CSSA, and SSSA, Madison, WI.

Durand, P., L. De Sa, A. Druilhet, and F. Said. 1991. Use of the inertial dissipation method for calculating turbulent fluxes from low-level airborne measurements. J. Atmos. Oceanic. Technol. 8:78–84.

Duxbury, J.M., D.R. Bouldin, R.E. Terry, and R.L. Tate. 1982. Emissions of nitrous oxide from soils. Nature (London) 298:462–464.

Fairall, C.W., and S.E. Larsen. 1986. Inertial dissipation methods and turbulent fluxes at the air ocean interface. Boundary-Layer Meteorol. 34:287–301.

Fan, S.M., S.C. Wofsy, P.S. Bakkwin, D.J. Jacob, S.M. Anderson, D.L. Kebabian, J.B. Mc-Manus, C.F. Kalb, and D.R. Fitzjarrald. 1992. Micrometeorological measurements of CH_4 and CO_2 exchange between the atmosphere and subarctic tundra. J. Geophys. Res. (In press.)

Fowler, D., and J. Duyzer. 1990. Micrometeorological techniques for the measurement of trace gas exchange. p. 189–208. In M.O. Andrea and D.S. Schimel (ed.) Exchange of trace gases between terrestrial ecosystems and the biosphere. Dahlem Konferenzen. John Wiley & Sons, Ltd., Chichester.

Gusten, H., G. Heinrick, R.W.H. Smidt, and U. Schurath. 1991. A novel ozone sensor for direct eddy flux measurements. J. Atmos. Chem. 14:73–84.

Hicks, B.B., D.D. Baldocchi, T.P. Meyers, R.P. Hosker, and D.R. Matt. 1987. A preliminary multiple resistance routine for deriving dry deposition velocities for measured quantities. Water Air Soil Pollut. 36:311–330.

Hicks, B.B., and R.T. McMillen. 1984. A simulation of the eddy accumulation method for measuring pollutant fluxes. J. Clim. Appl. Meteorol. 23:637–643.

Horst, T.W., and J.C. Weil. 1992. Footprint estimation for scalar flux measurements in the atmospheric surface layer. Boundary Layer Meteorol. 59:279–296.

Lenschow, D.H. 1986. Aircraft measurements in the boundary layer. p. 39–55. In D.H. Lenchow (ed.) Probing the atmospheric boundary layer. Am. Meteorol. Soc., Boston.

Lenschow, D.H., and B.B. Hicks. 1989. Global tropospheric chemistry. Rep. of the Workshop on Measurements of Surface Exchange and Flux Divergence of Chemical Species in the Global Atmosphere. NCAR, Boulder, CO.

Lenschow, D.H., E.R. Miller, and R.B. Friesen. 1991. A three-aircraft intercomparison of two types of air motion measurement systems. J. Atmos. Oceanic. Technol. 8:41–45.

Lenschow, D.H., R. Pearson, Jr., and B.B. Stankov. 1982. Measurements of ozone vertical flux to ocean and forest. J. Geophys. Res. 87:8833–8837.

MacPherson, J.I., and R.L. Desjardins. 1991. Airborne tests of flux measurement by the relaxed eddy accumulation technique. 7th Symp. on Meteorological Observations and Instrumentation, p. 6–11, New Orleans.

MacPherson, J.I., R. Grossman, and R.D. Kelly. 1992. Intercomparison results for FIFE flux aircraft. J. Geophys. Res. 97:18 499–18 514.

Mahrt, L., and W. Gibson. 1992. Flux decomposition into coherent structures. Boundary Layer Meteorol. 60:143–168.

Moore, C.J. 1986. Frequency response corrections for eddy correlation systems. Boundary Layer Meteorol. 37:17–35.

Mosier, A.R. 1989. Chamber and isotope techniques. p. 175–188. In M.O. Andrea and D.S. Schimel (ed.) Exchange of trace gases between terrestrial ecosystems and the biosphere. John Wiley and Sons Ltd., Chester, England.

Ogram, G.L., F.J. Northrup, and G.C. Edwards. 1988. Fast time response tunable diode laser measurements of atmospheric trace gases for eddy correlation. J. Atmos. Oceanic Technol. 5:521–527.

Pattey, E., R.L. Desjardins, F. Boudreau, and P. Rochette. 1992. Impact of density fluctuations on flux measurements of trace gases: Implications on the relaxed eddy accumulation technique. Boundary Layer Meteorol. 59:195–203.

Sachse, G.W., R.C. Harriss, L.O. Wade, and M.G. Perry. 1987. Fast-response, high-precision carbon monoxide sensor using a tunable diode laser absorption technique. J. Geophys. Res. 92:2071–2081.

Schuepp, P.H., M.Y. Leclerc, J.I. MacPherson, and R.L. Desjardins. 1990. Footprint prediction of scalar fluxes from analytical solutions of the diffusion equation. Boundary Layer Meteorol. 50:355–373.

Stewart, J.W.B., I. Aselmann, A.F. Bouwman, R.L. Desjardins, B.B. Hicks, P.A. Matson, H. Rodhe, D.S. Schimel, B.H. Svenson, R. Wassmann, M.J. Whiticar, and W.-X. Yang. 1989. Extrapolation of flux measurement to regional and global scales. p. 155–174. In M.O. Andrea and D.S. Schimel (ed.) Exchange of trace gases between terrestrial ecosystems and the biosphere. John Wiley and Sons Ltd., Chester, England.

Taylor, K.E., and M.C. MacCracken. 1990. Projected effects of increasing concentrations of carbon dioxide and trace gases on climate. p. 1–18. In B.A. Kimball et al. (ed.) Impact of carbon dioxide, trace gases, and climate change on global agriculture. ASA Spec. Publ. 53. ASA, Madison, WI.

Webb, E.K., G.I. Pearman, and R. Leuning. 1980. Correction of flux measurements for density effects due to heat and water vapour transfer. Q. J.R. Meteorol. Soc. 106:85–100.

Wesely, M.L. 1988. Use of variance techniques to measure dry air-surface exchange rates. Boundary Layer Meteorol. 44:13–31.

Wesely, M.L., J.A. Eastman, D.H. Stedman, and E.O. Yalvac. 1982. An eddy correlation measurement of NO_2 flux to vegetation and comparison to O_3 flux. Atmos. Environ. 16:815–820.

Wyngaard, J.C. 1988. Flow-distortion effects on scalar flux measurements in the surface layer: Implications for sensor design. Boundary Layer Meteorol. 42:19–26.

4 Use of Chamber Systems to Measure Trace Gas Fluxes

G.L. Hutchinson

USDA-ARS
Fort Collins, Colorado

G.P. Livingston

TGS Technology, Inc.
NASA Ames Research Center
Moffett Field, California

Trace gas exchange between soil–plant systems and the atmosphere is a complex phenomenon driven by a different set of physical, chemical, and biological processes for each chemical species and each environment (Fig. 4–1). For example, the exchange rate of a relatively inert gas like N_2O represents simply the difference between its integrated rates of production and consumption by soil biochemical processes. More reactive gases are subject to additional emission and deposition processes, such as foliar exchange (NH_3) or photochemical oxidation followed by deposition of the reaction products (NO). Further confounding the measurement and understanding of trace gas exchange across the surface–atmosphere boundary, the relative importances of the source and sink processes shown in Fig. 4–1 also vary with the rate of gas transport by diffusive, advective, and plant-mediated processes, and with the time and space scales over which the exchange is considered. Because of this complexity, great care must be exercised in drawing inferences about regional to global and seasonal to annual trace gas exchange rates from fluxes measured over small areas and short times.

Denmead and Raupach (1993) introduced the two most common approaches to field measurement of trace gas exchange—chamber methods and micrometeorology. Because tower-based micrometeorological techniques integrate the flux over a larger area (typically 10^2–10^3 m^2), they offer a potential advantage where the exchange rate is highly variable on the local scale measured by chamber methods (typically < 1 m^2). Chambers offer other advantages, including low cost and ease of use, and they represent the measurement technique of choice in process-level studies and other research requiring replicate measurements coincident in space or time. Our purpose is to provide sufficient insight into the use of chamber systems to enable poten-

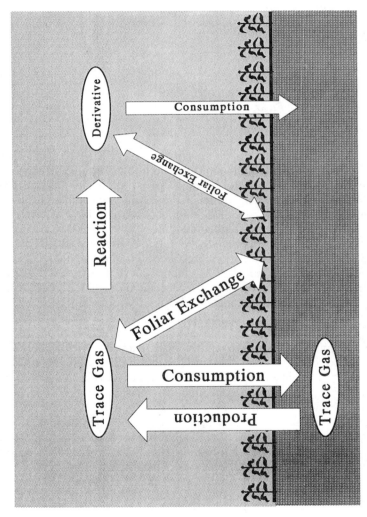

Fig. 4–1. Conceptual diagram of the major source and sink processes that contribute to trace gas exchange between soil–plant systems and the atmosphere.

tial users to identify applications where this approach is appropriate. We examine the most consequential sources of error in chamber-based flux measurements, suggest schemes for prioritizing and minimizing these errors, and then conclude by briefly discussing sampling design and data analysis strategies that can either reinforce advantages of the chamber approach, or alternatively, lead to serious misrepresentation of the measured exchange rates. Because of our experience, we focus on the measurement of soil gas exchange, but many of the same principles apply to measuring foliar gas exchange using either whole-plant or single-leaf chambers. The use of these chambers was recently reviewed by Reicosky (1990) and Field and Mooney (1990), respectively.

SOURCES OF ERROR IN INDIVIDUAL CHAMBER-BASED FLUX ESTIMATES

Following conventional terminology, open chambers are defined here as those with forced flow-through air circulation in which gases emitted by the soil are continually diluted and removed with air from outside the enclosure; closed chambers are those with closed-loop or no forced air exchange in which the emitted gases are allowed to accumulate. In either case, most users favor sealing the chamber to a permanently installed collar inserted into the soil, as opposed to inserting the chamber wall into the soil at the beginning of each measurement period (Mosier, 1989). The latter design risks opening direct channels to the soil surface for air that could otherwise escape to the atmosphere only along a tortuous diffusion pathway. Although a permanent collar may have a long-term influence on the microclimate of the site from which soil gas exchange is to be measured, this effect is probably small if the collar is constructed from heat-insulating material resistant to microbial decomposition, and its height above the soil surface is minimized.

Let us assume that the measured flux, f, whether based on open- or closed-chamber concentrations, can be represented by the linear combination

$$f = \phi + \delta + \epsilon \qquad [1]$$

where the parameter ϕ is the true flux (in units of mass area^{-1} time^{-1}), δ represents all sources of bias, and ϵ all sources of random variability. The two types of error in Eq. [1] were given different symbols to emphasize the large, and often unrecognized, difference between them. Random error (ϵ) is a measure of precision, i.e., the repeatability of replicate measurements. Its magnitude can be estimated by basic sampling principles and statistical measures such as the variance, or average squared deviation of replicate observations from the mean. In contrast, bias is the magnitude and direction of the tendency to measure something other than what was intended. There are no measures of bias, except in the uncommon situation where ϕ is known. In the following discussion, we have grouped potential sources of bias into two broad categories: (i) physical and biological disturbances associated with the measurement process, and (ii) errors associated with sample handling, sample analysis, and inaccurate models or inappropriate methods for computing flux from measured concentration data. We will argue that although they represent potentially substantial sources of error, both categories can be minimized or eliminated by prudent chamber design and strict adherence to a carefully conceived measurement protocol, thus leading to the conclusion that the existence of significant bias in an experimental dataset represents user error.

Physical and Biological Disturbances

Both open- and closed-chamber systems have been criticized as susceptible to bias resulting from physical and biological disturbances associated with the measurement process (e.g., Cropper et al., 1985; Denmead, 1979; Knapp & Yavitt, 1992; Nakayama, 1990). The bias results from perturbations of environmental parameters that regulate the processes shown in Fig. 4–1, as well as from creation of the potential for additional source or sink processes. Several such sources of bias are described in the following paragraphs.

Temperature Effects

When either an open or closed chamber is deployed in the field, there is opportunity for temperatures inside and outside the chamber to differ, with resulting potential impact on trace gas production, consumption, and transport processes in the covered soil (Fig. 4–2). The magnitude of this per-

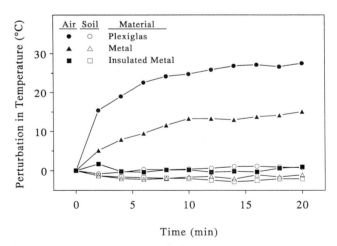

Fig. 4–2. Differences in air temperatures measured 10 cm above the soil surface and soil temperatures measured 2 cm below the soil surface inside and outside chambers (88-cm diam., 15-cm height) constructed from Plexiglas, metal, or insulated metal (data from Matthias et al., 1980).

turbation can be minimized by adopting reasonable precautions to minimize solar heating of the chamber, such as insulating the chamber walls and applying a reflective covering to their external surfaces (Hutchinson & Mosier, 1981), or by employing an external temperature controller (Bartlett et al., 1990). In addition, soil temperature follows air temperature only after a significant lag that increases with soil depth, so the effect of air temperature perturbations is likely minimal over relatively short deployment times.

Pressure Effects

Two types of pressure disturbances can be generated by the use of chamber systems—those that result in a difference between mean air pressure inside and outside the enclosure, and those that subdue pressure fluctuations associated with the turbulence in air movement over the soil surface. The first type is of primary concern with open chambers, which must necessarily be operated at pressures above or below the ambient level to establish flow-through air circulation, thereby creating the potential for transport of soil gases by mass flow (Cropper et al., 1985; Kanemasu et al., 1974). If the chamber's air inlet is properly sized and configured, this potential source of bias is probably negligible (Denmead, 1979). A properly vented closed chamber causes no perturbation in mean air pressure except that potentially associated with its installation (Hutchinson & Mosier, 1981).

Transmitting high frequency air pressure fluctuations to the covered soil surface is a more difficult problem, but one of particular importance during measurement of low trace gas exchange rates over porous soils (Fig. 4-3). If the pumping action imposed on surface layers of soil by turbulence-induced pressure fluctuations is eliminated, exchange of the measured trace gas may

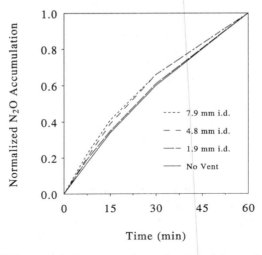

Fig. 4-3. Mean N_2O accumulation expressed as a fraction of the total accumulated in 1 h beneath closed chambers with different vent tube diameters; optimum vent inner diameter was computed to be 5 mm from equations proposed by Hutchinson and Mosier (1981) (data from Hutchinson & Mosier, 1981).

be significantly reduced until its soil concentration gradient adjusts to a higher level capable of again driving a flux equal to the rate of net production (Kimball, 1983; Kimball & Lemon, 1972). Simply drilling a hole in the cover wall allows ambient pressure fluctuations to reach the covered soil, but this solution also permits the exchange of ambient and enclosed air. To overcome both problems, Hutchinson and Mosier (1981) proposed that pressure communication occur through an open tube with internal volume large enough that air exhausted from the chamber during a decline in ambient pressure is captured within the tube and then returned to the chamber when the pressure rises again; loss of the measured trace gas is limited in this case to that which diffuses down the tube in response to the difference in concentration across the chamber wall. The same authors also developed equations for estimating optimum vent tube diameter and length as a function of chamber and environmental parameters.

Concentration Effects

The exchange rate of a trace gas across the soil–atmosphere boundary is largely a function of its diffusion coefficient and concentration gradient between sites of production (or consumption) and the soil surface. Unfortunately, that gradient begins to diminish immediately following chamber deployment. For open chambers, where the exchange rate is computed from the difference between inlet and outlet concentrations of the gas under the assumption that its gradient in the underlying soil is at steady state, the true exchange rate may be significantly underestimated unless sufficient time is allowed for the gradient to adjust to the perturbed concentration in the chamber atmosphere. Jury et al. (1982) showed by simulation that this time may be quite long in some cases and is rather poorly defined in most.

Beneath a closed chamber, perturbation of the exchange rate continues throughout its deployment, causing continual declines in both the soil gradient and the rate of change in trace gas concentration of chamber air. As a result, the equation most often used to compute the exchange rate from observed closed chamber concentrations necessarily yields an underestimate, because it assumes a linear change in trace gas concentration C (and therefore, constant flux) over time t; i.e.,

$$f = \frac{V(C_t - C_0)}{At} \qquad [2]$$

where V is the internal volume of the chamber, and A is the soil area that it covers (e.g., Nakayama, 1990). To compensate, most users attempt to sample over time periods short enough that C can be reasonably approximated as a linear function of t. In many situations, however, particularly during measurement of small exchange rates from porous soils, the limits imposed by sampling logistics and analytical precision preclude sufficient reduction in length of the sampling period to avoid significant nonlinearity (Anthony & Hutchinson, 1990).

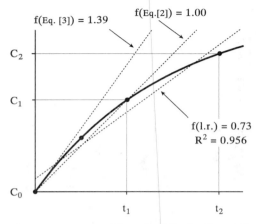

Fig. 4-4. Trace gas exchange rates computed from Eq. [2] using C_0 and C_1, from Eq. [3] using C_0, C_1, and C_2, and from linear regression analysis (l.r.) using all four data points of a hypothetical nonlinear dataset defined by the criteria that it meet the assumptions on which Eq. [3] is based and that $(C_2 - C_1) = 0.5 (C_1 - C_0)$; arbitrary units define the Eq. [2] flux computed at t_1 to be unity.

To address this issue, Matthias et al. (1978) and Hutchinson and Mosier (1981) proposed nonlinear models based on diffusion theory. The first of these requires an iterative solution, which makes it difficult to apply, but the latter (Eq. [3]) is easily implemented if the change in chamber concentrations is measured over two successive time periods of equal length.

$$f = \frac{V (C_1 - C_0)^2}{At(2C_1 - C_2 - C_0)} \ln \left[\frac{C_1 - C_0}{C_2 - C_1} \right] \text{ for } t_2 = 2t_1$$

$$\text{and } \frac{C_1 - C_0}{C_2 - C_1} > 1 \qquad\qquad [3]$$

The large difference in exchange rates estimated by linear vs. nonlinear models is illustrated in Fig. 4-4, where it is easy to visualize that Eq. [3] estimates the flux (or slope of the trace gas accumulation curve) at time zero, when the chamber has minimum impact on the flux it was intended to measure. Discussion of this issue is continued in the section concerning flux estimation.

Site Disturbances

In addition to the environmental perturbations just described, use of chamber systems may also cause site disturbances that have equal or even greater potential for altering the measured flux. For example, soil compaction may occur within or around the chamber, causing horizontal gas transport that invalidates the assumed correspondence of the observed trace gas flux with net production by the covered soil. Compaction within a perma-

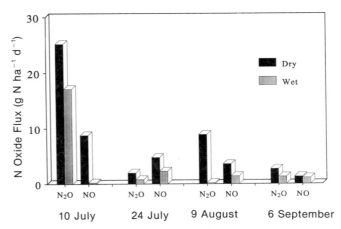

Fig. 4–5. Mean NO and N_2O fluxes from irrigated (wet) and nonirrigated (dry) furrows in a corn field 1 h after infiltration of the water applied to alternate furrows (data from Guenzi et al., 1990).

nent collar is most likely to occur during its installation, while that outside the collar results from foot traffic necessary to service the chamber during periodic flux measurements. The latter may become increasingly more serious in long-term studies unless it is circumvented by using elevated walkways or other precautions. Similar bias due to horizontal gas transport may result from applying artificial rainfall to only the covered soil, or from excluding natural precipitation. For example, Guenzi et al. (1990) demonstrated that following irrigation of alternate furrows in a field of corn (*Zea mays* L.), NO and N_2O produced beneath the saturated soil moved laterally and then escaped to the atmosphere from adjacent nonirrigated furrows where the surface soil had less restriction to gas transport (Fig. 4–5).

Some biological disturbance may also occur during trace gas exchange measurement by chamber methods. For example, inserting a permanent collar into the soil causes substantial root damage in many ecosystems, influencing not only plant nutrient uptake, but also nutrient availability to the soil microbial community (Matson et al., 1990). To compensate, the collars should be installed at least several days prior to data collection to allow time for the effects of this disturbance to subside. Because plants sometimes serve as a conduit for transport of gases between soil and the atmosphere (particularly in wetland ecosystems), and because exudates from plant wounds represent potential substrate for microbial growth, damage to aboveground plant parts must also be minimized. Although photosynthesis, transpiration, and selected other plant and microbial physiological processes probably respond immediately to artificial environmental conditions imposed by the chamber (e.g., darkness, elevated CO_2 and water vapor concentrations), the short-term impact of these changes on the soil exchange rates of most trace gases is likely to be minimal.

Other physical and biological disturbances may also be associated with the use of chamber systems, but they, too, likely can be controlled by using appropriate chamber design, relatively short deployment times, and reasonable user care to minimize site disturbances. We conclude that susceptibility to such disturbances does not seriously limit the successful application of chamber methods.

Concentration Measurement and Flux Estimation

Sample Handling and Analytical Bias

Additional bias can be introduced to chamber-based estimates of trace gas exchange during collection, transport, storage, and analysis of samples of the enclosed air. The analyst must remain constantly alert, asking "Is there opportunity for reaction with or sorption by the walls of containers or tubing used to transport air samples from the chamber to an appropriate analytical device? For sample loss due to leakage? For dissolution of the trace gas in water that might condense from the sample if it is cooled? For chemical, biological, or photochemical production or consumption of the trace gas during sample transport or storage? Etc." Although these common sources of bias can be minimized or eliminated by adopting appropriate precautions, even experienced analysts encounter situations where less obvious sources of sample handling bias result in faulty estimates of trace gas exchange rates. The best protection against such situations is careful comparison of the concentration and flux data with ancillary environmental parameters or concurrent measures of other trace gas concentrations and exchange rates.

Sources of bias associated with the analytical determination of trace gas concentration depend on both the properties of the gas and the technique chosen for its analysis (e.g., Edwards & Sollins, 1973). The potential for bias from this source is again substantial, but like sample handling bias, it is not unique to chamber-based measurement techniques and can usually be nearly eliminated by meticulously following prudent analytical protocol. Examples of analytical bias and the difficulties it creates appear in the following section.

Model and Estimation Bias

A major, but frequently overlooked, source of bias in individual chamber-based estimates of trace gas exchange is use of inappropriate procedures for computing flux from the observed concentration data. Probably the most serious of these procedural errors is failure to account for feedback effects of the chamber on the exchange process it was intended to measure. As implied previously, these effects can be minimized for closed chambers by using the observed data to predict the exchange rate for that moment when the chamber's impact is minimal, i.e., at the time when it was installed. The procedure for making this prediction is formally defined by statistical principles and encompasses: (i) defining a model of the exchange process, (ii) verifying the model's fit to the observed concentration data, (iii) predicting the unperturbed exchange rate at $t = 0$, and (iv) testing the significance of the prediction. Existing trace gas exchange literature gives little attention to the steps involving model validation and significance testing, but because considerable bias can be introduced by their omission, all four steps are essential to establishing the credibility of exchange rate estimates. The required statistics are easily computed by the spreadsheet or statistics software available in nearly every laboratory, so the effort involved is minimal.

Numerous models can be defined to relate the exchange rate of a trace gas to its time-dependent concentration in a closed-chamber atmosphere. The choice should be based on the criteria that the model must represent the exchange process, its goodness of fit to the observed concentration data can be verified, and the significance of the resulting flux estimate can be tested. Moreover, because there is no reason to assume that the same model should apply throughout any experiment, the choice may vary from chamber to chamber, particularly if the sampling environment is heterogeneous.

In those situations where the chamber has minimal impact on the measured exchange rate, a simple linear regression model is usually applicable. This approach is equivalent to adopting Eq. [2], but generalizing it to take advantage of multiple concentration observations. Linear regression can not be indiscriminately applied in all situations, however. A nonlinear model must often be employed to account for the diminishing rate of change in the trace gas concentration of chamber air as a function of time. Application of a linear model in such cases may seriously underestimate the unperturbed rate of trace gas exchange. For example, when linear regression is used to establish the slope of a line through the four data points in Fig. 4–4, the resulting estimate of f is 0.73, only 53% of the true flux (ϕ) defined for this hypothetical dataset. It is especially noteworthy that this large underestimation of f occurs despite the appearance of relatively small deviation from linearity (i.e., $R^2 = 0.956$). Inappropriate application of a linear model to nonlinear data may also increase the coefficient of variation in the flux estimates, because the amount of introduced bias depends on the degree of nonlinearity in the measured trace gas accumulation rate, which varies from chamber to chamber.

Whether employing simple linear regression or a nonlinear exchange model, it is particularly important to verify that assumptions on which the model is based are satisfied (Steel & Torrie, 1960; Dowdy & Wearden, 1983). The assumption most subject to violation concerns the model's fit to the observed data. Nonlinear models can be fit either directly or by appropriately transforming the data such that linear regression can be subsequently employed. The latter approach offers the advantage of model verification and significance testing using readily available statistical tools. Models based on diffusion theory such as those proposed by Matthias et al. (1978) and Hutchinson and Mosier (1981) offer the advantage of a theoretical construct and with further development may prove amenable to validation and significance testing. Usually, a careful examination of appropriate summary statistics and a scatterplot of actual vs. predicted concentrations as a function of time are the most efficient means of judging model fit in trace gas exchange studies where the number of observations is generally limited. Other potential violations of model-fitting assumptions that pose less serious difficulties include: (i) lack of independence in the deviations between actual and predicted concentrations over time, which might arise, for example, from sample gain or loss between the times of collection and analysis, or from uncorrected instrument drift; and (ii) inequality in the variances of the concentration measures over time, which is usually of concern only where measurement precision is a function of concentration.

After model validation, the predicted relation between concentration and time must be tested at a preselected level of significance. For linear models, the null hypothesis (H_0) is that the slope of the regression line is equal to zero; for nonlinear models, an analogous test must be defined. In either case, the appropriate statistic is a t test defined by the ratio of the estimated exchange rate to the standard error (SE) of the estimate at a $1 - \alpha$ significance level and $n - 2$ degrees of freedom, where n is the number of concentration measurements. Estimates of trace gas exchange rates should be ruled valid only when H_0 is rejected at the preselected significance level (nominally, $\alpha = 0.05-0.10$); if H_0 can not be rejected at the given level of precision of the measurement process, then the best estimate of the exchange rate over the sampling period is zero.

The procedure outlined above not only encourages greater rigor by minimizing bias in the estimation of trace gas exchange rates from observed concentration data, but also facilitates identification and prioritization of sources of imprecision, thus providing the information needed to design measurement protocols with improved efficiency and sensitivity. Existing trace gas exchange literature gives little attention to estimating measurement precision. The "minimum detectable flux" reported by some authors is based on overall measurement precision, while in other cases it is based solely on analytical precision. We believe a definition based on the observed exchange rate and its SE for each chamber is the most rigorous and useful, although it is a liberal definition which presumes that variability associated with the concentration measurements is the only source of imprecision in the flux estimates. Using terms defined above

$$\text{minimum detectable flux} = (t)(SE) \qquad [4]$$

for $n - 2$ degrees of freedom and a $1 - \alpha$ significance level. Summaries of this parameter, using all or any subset of the estimates from an experiment, can then be prepared using descriptive statistics.

General guidelines for improving sensitivity of the measurement protocol that can be derived from Eq. [4] are: (i) the time period over which samples are collected should be maximized within constraints imposed by model assumptions, (ii) the number of concentration determinations for each flux estimate should also be maximized, and (iii) sample handling variance should be minimized, while maximizing the measurement precision for all parameters in the flux model. Increasing length of the sampling period reduces SE, but must not compromise model assumptions regarding the chamber's impact on the exchange process. Increasing n provides even greater reduction in minimum detectable flux through both qualitative means (due to improved ability to evaluate whether the data meet model assumptions) and quantitative means (due to the effect of sample size on both terms in Eq. [4]). Using linear regression analysis as an example, the gain in detectability realized by increasing n from three to four temporally independent observations is nearly threefold (with all other factors equal). Although the return quickly diminishes thereafter, an additional 1.4-fold gain is realized by increasing n from four to five. Replicate observations at specific times also contribute enhanced sensitivity, but unless linearity can be otherwise justified, maximum gain is achieved by distributing observations throughout the sampling period. Analogous arguments apply to nonlinear flux models.

The third guideline reiterates the need to minimize potentially large errors associated with collection, transport, storage, and analysis of chamber air samples, as well as those associated with the measurement of other parameters in the flux model (V and A), which may be particularly difficult when the soil surface is not level or when a significant volume of plant material is enclosed by the chamber. Beyond a certain level of performance, however, further increases in the measurement precision of individual model parameters can be achieved only at substantial cost in time or expense, which must be weighed against the cost of enhancing the overall detection limit by increasing sampling effectiveness.

SAMPLING DESIGN AND DATA ANALYSIS ERRORS

Although potential sources of bias in individual chamber-based estimates of trace gas exchange are numerous, their combined effect may be dwarfed by the bias injected into an experimental dataset as a whole by using a sampling design inappropriate for addressing the intended research objectives, or by performing subsequent data analyses that fail to consider the original sample allocation. As a consequence of these errors, the frequency distribution, magnitude, and variability associated with multiple exchange rate observations combined over space and time may be significantly misrepresented. Sampling theory based on random allocations is well established yielding both parametric and nonparametric analytical methods (e.g., Cochran & Cox, 1957; Conover, 1971), and nonrandom sampling strategies are developing that meet other objectives (Robertson, 1987; Isaaks & Srivastava, 1989), but misuse of sampling designs and analyses remains a major issue in ecological studies (Hurlbert, 1984; Day & Quinn, 1989). Moreover, because the database of exchange rate measurements for some trace gases is so limited, our understanding of the science issues, as well as funding and policy decisions, are particularly sensitive to sampling biases.

Appropriate allocation of the number and location of chambers must consider not only the factors regulating the trace gas exchange rate, but also the requirements and assumptions of the anticipated statistical analyses. As a consequence, process level and regional survey objectives can rarely be properly addressed by a single sampling design. Field studies are particularly susceptible to confoundment, because the exchange rate controllers may vary along different environmental gradients. A sampling design proposed to assess the effect of one environmental factor must, therefore, be carefully randomized to minimize confoundment by another.

On the other hand, logistics and economics often require the intentional confoundment of selected sources of variance to increase the number of obervations sufficiently that the primary objective can be tested with reasonable sensitivity. The necessary cost of such approaches is that individual contributions to the confounded variance can no longer be evaluated. For example, temporal variability attributable to hourly and daily changes in environmental conditions over a given period may be combined with the spatial variability of the study area to yield an overall variance against which regional scale hypotheses may be tested. Conversely, multiple observations on local scales of a few meters may provide suitable information for evaluating local environmental controls, but rarely meet the spatial independence criterion required for them to be treated as replicate observations on a regional basis.

An often-heard critique of chamber-based studies concerns the difficulty of allocating a sufficient number of observations to "significantly" characterize the trace gas exchange rate of a particular region because of the tremendous variability encountered over local temporal and spatial scales. The exchange rates of some gases differ by one to two orders of magnitude across distances less than a meter (Folorunso & Rolston, 1984; Morrissey & Livingston, 1992) and times less than an hour (Williams & Fehsenfeld, 1991), which is likely the same magnitude of variability observed on regional or seasonal scales. As a result, the probability distributions of chamber-based flux estimates are frequently log-normal or at least strongly positively skewed, and usually truncated, or noncontinuous below zero. In most cases, however, this issue can be effectively addressed by employing stratified sampling approaches (i.e., grouping the area or time of study into strata hypothesized to have similar exchange rates) that permit subsets of the resulting exchange rate estimates to be weighted by the areal and temporal extent over which they apply (Green, 1979).

SUMMARY AND RECOMMENDATIONS

Trace gas exchange between soil–plant systems and the atmosphere is a complex phenomenon, whose measurement by chamber systems is subject to many potential sources of bias and random variability. However, bias due to physical and biological disturbances associated with the measurement process can be mostly overcome by using appropriate chamber design, relatively short deployment times, and reasonable care to minimize site disturbances. Similarly, bias introduced during trace gas concentration measurement and flux estimation can be minimized by following a carefully conceived measurement protocol and statistically sound procedures for choosing an appropriate model to estimate flux from the observed concentrations. To a large extent, then, the existence of significant bias in trace gas exchange estimates represents user error. Random error is also partially under user control, and its magnitude can be reduced by increasing the number of observations.

Although potential sources of error in individual chamber-based estimates of trace gas exchange are numerous, the use of sampling designs inappropriate for addressing the intended research objectives and subsequent analyses that fail to consider the original sample allocation probably represent the most serious critiques of chamber-based studies. Current high interest in developing both process level and global/annual scale understanding of trace gas exchange magnifies concern about misrepresentation of data by either the original research report or its readers. We urge authors, reviewers, and editors to insist on explicit and complete identification of assumptions inherent in sampling designs and data applications.

The choice of flux measurement technique must always be judged against the principal criterion that it best supports the research objectives. Although micrometeorological techniques offer an advantage in some research with primary objective to estimate regional exchange rates, chamber systems represent the measurement technique of choice in research with objectives requiring temporal or spatial replication of the measured flux with manageable cost and logistics. Trace gas concentrations and environmental conditions in a carefully designed open chamber may remain relatively close to those in situ, an important advantage in situations requiring continuous monitoring over periods exceeding a few tens of minutes (Matthias et al., 1978; Denmead, 1979). Whenever the research objectives can be achieved by periodic "instantaneous" flux measurements, however, a closed chamber may be a better choice (Hutchinson & Mosier, 1981; Sebacher & Harriss, 1982), because: (i) gases emitted by the soil are not continually diluted with external air, so smaller exchange rates can be measured at any given level of analytical precision, (ii) the typically shorter deployment times of closed chambers should result in smaller disturbance of the microsite under study, and (iii) models for predicting the trace gas exchange rate at the time of closed chamber installation are more amenable to rigorous validation and significance testing.

REFERENCES

Anthony, W.H., and G.L. Hutchinson. 1990. Soil cover measurement of N_2O flux: Linear vs. nonlinear equations. p. 243. *In* Agronomy abstracts. ASA, Madison, WI.

Bartlett, D.S., G.J. Whiting, and J.M. Hartman. 1990. Use of vegetation indices to estimate intercepted solar radiation and net carbon dioxide exchange of a grass canopy. Remote Sens. Environ. 30:115–128.

Cochran, W.G., and G.M. Cox. 1957. Experimental designs. John Wiley & Sons, New York.

Conover, W.J. 1971. Practical nonparametric statistics. 2nd ed. John Wiley & Sons, New York.

Cropper, W.P., Jr., K.C. Ewel, and J.W. Raich. 1985. The measurement of soil CO_2 evolution *in situ*. Pedobiologia 28:35–40.

Day, R.W., and G.P. Quinn. 1989. Comparison of treatments after an analysis of variance in ecology. Ecol. Monogr. 59:433–463.

Denmead, O.T. 1979. Chamber systems for measuring nitrous oxide emission from soils in the field. Soil Sci. Soc. Am. J. 43:89–95

Denmead, O.T., and M.R. Raupach. 1993. Methods for measuring atmospheric gas transport in agricultural and forest systems. p. 19–43. *In* L.A. Harper et al. (ed.) Agricultural ecosystem effects on trace gases and global climate change. ASA Spec. Publ. no. 55. ASA, CSSA, SSSA, Madison, WI.

Dowdy, S., and S. Wearden. 1983. Statistics for research. John Wiley & Sons, New York.

Edwards, N.T., and P. Sollins. 1973. Continuous measurement of carbon dioxide evolution from partitioned forest floor components. Ecology 54:406–412.

Field, C.B., and H.A. Mooney. 1990. Leaf chamber methods for measuring photosynthesis under field conditions. p. 117–139. *In* N.S. Goel and J.M. Norman (ed.) Remote sensing reviews, Vol. 5. Issue 1. Instrumentation for studying vegetation canopies for remote sensing in optical and thermal infrared regions. Harwood Acad. Publ., New York.

Folorunso, O.A., and D.E. Rolston. 1984. Spatial variability of field-measured denitrification gas fluxes. Soil Sci. Soc. Am. J. 48:1214–1219.

Green, R.H. 1979. Sampling design and statistical methods for environmental biologists. John Wiley & Sons, New York.

Guenzi, W.D., G.L. Hutchinson, and W.E. Beard. 1990. Nitric and nitrous oxide emissions and soil nitrate distribution in a center-pivot-irrigated cornfield. p. 250. *In* Agronomy abstracts. ASA, Madison, WI.

Hurlbert, S.H. 1984. Pseudoreplication and the design of ecological field experiments. Ecol. Monogr. 54:187–211.

Hutchinson, G.L., and A.R. Mosier. 1981. Improved soil cover method for field measurement of nitrous oxide fluxes. Soil Sci. Soc. Am. J. 45:311–316.

Isaaks, E.H., and R.M. Srivastava. 1989. An introduction to applied geostatistics. Oxford Univ. Press, New York.

Jury, W.A., J. Letey, and T. Collins. 1982. Analysis of chamber methods used for measuring nitrous oxide production in the field. Soil Sci. Soc. Am. J. 46:250–256.

Kanemasu, E.T., W.L. Powers, and J.W. Sij. 1974. Field chamber measurements of CO_2 flux from soil surface. Soil Sci. 118:233–237.

Kimball, B.A. 1983. Canopy gas exchange: Gas exchange with soil. p. 215–226. *In* H.M. Taylor et al. (ed.) Limitations to efficient water use in crop production. ASA, SCSSA, SSSA, Madison, WI.

Kimball, B.A., and E.R. Lemon. 1972. Theory of soil air movement due to pressure fluctuations. Agric. Meteorol. 9:163–181.

Knapp, A.K., and J.B. Yavitt. 1992. Evaluation of a closed-chamber method for estimating methane emissions from aquatic plants. Tellus 44B:63–71.

Matson, P.A., P.M. Vitousek, G.P. Livingston, and N.A. Swanberg. 1990. Sources of variation in nitrous oxide flux from Amazonian ecosystems. J. Geophys. Res. 95:16 789–16 798.

Matthias, A.D., A.M. Blackmer, and J.M. Bremner. 1980. A simple chamber for field measurement of emissions of nitrous oxide from soils. J. Environ. Qual. 9:251–256.

Matthias, A.D., D.N. Yarger, and R.S. Weinbeck. 1978. A numerical evaluation of chamber methods for determining gas fluxes. Geophys. Res. Lett. 5:765–768.

Morrissey, L.A., and G.P. Livingston. 1992. Methane emissions from Alaska arctic tundra: An assessment of local spatial variability. J. Geophys. Res. 97:16 661–16 670.

Mosier, A.R. 1989. Chamber and isotope techniques. p. 175–187. *In* M.O. Andreae and D.S. Schimel (ed.) Exchange of trace gases between terrestrial ecosystems and the atmosphere. John Wiley & Sons, Chichester.

Nakayama, F.S. 1990. Soil respiration. p. 311–321. *In* N.S. Goel and J.M. Norman (ed.) Remote sensing reviews. Vol. 5. Issue 1. Instrumentation for studying vegetation canopies for remote sensing in optical and thermal infrared regions. Harwood Acad. Publ., New York.

Reicosky, D.C. 1990. Canopy gas exchange in the field: Closed chambers. p. 163–177. *In* N.S. Goel and J.M. Norman (ed.) Remote sensing reviews. Vol. 5. Issue 1. Instrumentation for studying vegetation canopies for remote sensing in optical and thermal infrared regions. Harwood Acad. Publ., New York.

Robertson, P. 1987. Geostatistics in ecology: Interpolation with known variance. Ecology 68:744–748.

Sebacher, D.I., and R.C. Harriss. 1982. A system for measuring methane fluxes from inland and coastal wetland environments. J. Environ. Qual. 11:34–37.

Steel, R.G.D., and J.H. Torrie. 1960. Principles and procedures of statistics. McGraw Hill, New York.

Williams, E.J., and F.C. Fehsenfeld. 1991. Measurement of soil nitrogen oxide emissions at three North American ecosystems. J. Geophys. Res. 96:1033–1042.

5 Processes for Production and Consumption of Gaseous Nitrogen Oxides in Soil

G.L. Hutchinson

USDA-ARS
Fort Collins, Colorado

Eric A. Davidson

Woods Hole Research Center
Woods Hole, Massachusetts

Gaseous N oxides, N_2O and NO_x (NO + NO_2), are trace atmospheric constituents that function directly or indirectly as potentially important greenhouse gases in various global climate change scenarios (Duxbury et al., 1993). Both also participate in the production and/or consumption of atmospheric oxidants (e.g., O_3, OH), and NO_x is removed from the atmosphere in a series of photochemical reactions that result in formation of HNO_3, the fastest-growing component of acidic deposition (Logan, 1983). In addition to these important impacts on the chemistry of the atmosphere, it has been suggested that soil emission of NO_x may comprise a significant fraction of the unaccounted losses typically observed in the N balance sheets of fertilized agricultural soils, and that the emission, transport, and subsequent redeposition of NO_x results in substantial N redistribution both within and among natural and agricultural ecosystems (Williams et al., 1992). Because biochemical processes in soil are some of the principal sources of atmospheric N oxides, it then becomes important to understand these processes and to identify important physical, chemical, and biological controllers of the process rates. In this paper we review existing knowledge of these processes and their controllers at the cellular level, and then comment briefly on the challenges associated with applying that knowledge at field and landscape scales. Field and laboratory measurements of actual soil NO and N_2O exchange rates are included only to the extent needed to illustrate the responsible processes; a comprehensive review of these data was recently completed by Williams et al. (1992).

CELLULAR LEVEL CONTROL OF SOIL NITRIC OXIDE
AND NITROUS OXIDE EXCHANGE

Both biotic and abiotic processes are involved in the production of NO_x and N_2O in soils. For both process types, it is generally agreed that the preponderance of NO_x emitted by soil is NO, with direct soil emission of NO_2 accounting for substantially less than 10% of the total. Instances where a larger fraction has been reported as NO_2 can probably be attributed either to postemission oxidation of NO prior to sample analysis or to nonspecificity of the technique chosen for determining the NO_2 fraction of NO_x (Williams et al., 1992). Among biotic sources, numerous groups of soil microorganisms contribute to the production of NO and N_2O through a variety of biochemical reactions. The bacterial processes of nitrification and denitrification are generally accepted to be the principal sources in soil, but most microbial processes that involve oxidation or reduction of N through the +1 or +2 oxidation state probably yield at least trace amounts of the two gases (Firestone & Davidson, 1989). For example, N_2O production by processes other than nitrification or denitrification has been observed in acid forest soils (Robertson & Tiedje, 1987) and in pure cultures of some fungi (Bollag & Tung, 1972; Burth & Ottow, 1983), so a fungal source seems likely. Nitrous oxide loss also occurs during assimilatory NO_3^- reduction by some yeasts (Bleakley & Tiedje, 1982), and both NO and N_2O are reportedly produced in varying amounts by nondenitrifying NO_3^--reducing bacteria (Smith & Zimmerman, 1981; Anderson & Levine, 1986). The importance of these and other nonnitrifying–nondenitrifying processes to total biogenic NO and N_2O production is uncertain; this topic was recently reviewed by Tiedje (1988).

Abiotic production of N_2O, and particularly NO, occurs primarily through a set of reactions collectively termed chemodenitrification. The most important of these reactions is the disproportionation of nitrous acid (HNO_2) known to occur in acid soils (Nelson, 1982), especially those high in organic matter content (Blackmer and Cerrato, 1986). Although this reaction has not been demonstrated in neutral or alkaline soils in the laboratory, its occurrence in the natural environment cannot be discounted because of the possible existence in undisturbed soils of microsites where the required accumulation of NO_2^- and low pH can occur as a result of solute concentration in thin water films during freezing or drying, or because of proximity to a colony of NH_4^+ oxidizers, etc. (Davidson, 1992a). McKenney et al. (1990) recently demonstrated that the stoichiometric NO yield of HNO_2 disproportionation was increased threefold by coupling it to reaction with organic nitrites in acid peat. Other abiological processes for NO and N_2O production in soil include decomposition of hydroxylamine (NH_2OH) and reaction of NO_2^- with the phenolic constituents of soil organic matter, among others (Nelson, 1982). The contribution of the latter reactions to soil emission of NO and N_2O is apparently surpassed by that of HNO_2 disproportionation, which itself makes a much smaller contribution than the biological processes of nitrification and denitrification discussed below.

Nitrification

Nitrification is defined as the biological oxidation of NH_4^+ to NO_2^- and NO_3^-, or a biologically induced increase in the oxidation state of N (Soil Science Society of America, 1987). Numerous studies have indicated that the process is a quantitatively important component of the N cycle in most cultivated agricultural soils. In contrast, nitrifying microorganisms were generally thought to have minor influence on N cycling in many mature natural ecosystems, because plants and heterotrophic microorganisms were believed to be better competitors for NH_4^+ (Bormann & Likens, 1979; Melillo, 1981). Recent evidence indicates, however, that the N cycles of at least some mature natural ecosystems include substantial nitrification activity. For example, Davidson et al. (1990) reported that between 12 and 46% of the N mineralized in an undisturbed N-limited California grassland was oxidized to NO_3^-, even during seasons when plant uptake and microbial immobilization were both active. They attributed the competitive success of the nitrifiers to microsite heterogeneity in the rates of NH_4^+ mineralization and of NH_4^+ assimilation by microorganisms and roots.

The process of nitrification is associated with the metabolism of chemoautotrophic bacteria of the family Nitrobacteriaceae, as well as several species of heterotrophic microorganisms. Heterotrophs such as *Aspergillus flavus* and *Alcaligenes* sp., among many others, have been reported to form NO_2^- or NO_3^- from NH_4^+ or other reduced forms of N when grown in culture media (Castignetti & Gunner, 1980; Schmidt, 1982). However, convincing evidence that relates the occurrence of a particular heterotroph in its natural environment to the progression of nitrification in that environment is limited to soils with pH too low or temperature too high to support the growth of chemoautotrophic nitrifiers (Stroo et al., 1986; Schimel et al., 1984).

It is widely accepted that the preponderance of nitrification in soil is accomplished by a few genera of chemoautotrophic bacteria—*Nitrosomonas* and *Nitrospira*, which oxidize NH_4^+ to NO_2^-, and *Nitrobacter*, which converts NO_2^- to NO_3^-. Although low numbers of a few other NH_4^+-oxidizing chemoautotrophs are also present in many soils, *Nitrobacter* is the only genus known to be involved in the oxidation of NO_2^-, so it is surprising that this ion is promptly oxidized and rarely accumulates in soil. Notable exceptions include tropical savannah (Johansson & Sanhueza, 1988) and seasonally dry tropical forest soils (Davidson et al., 1991) that accumulate NO_2^- during the dry season. Oxygen is obligatory for the chemoautotrophic oxidation of either NH_4^+ or NO_2^-, and both reactions are coupled to electron transport phosphorylation, thereby providing the energy required for growth and regeneration of the responsible organisms. All members of the family Nitrobacteriaceae are aerobes that synthesize their cell constituents from CO_2 by way of the Calvin reductive pentose phosphate cycle. The relatively narrow species diversity of the chemoautotrophs responsible for nitrification in soil render the process unusually susceptible to external influences (Haynes, 1986).

The biochemical pathway of chemoautotrophic nitrification remains a subject of much debate. There is good evidence that NH_2OH (N oxidation state -1) is the first intermediate product of NH_4^+ oxidation, but subsequent intermediates with N oxidation states $+1$ and $+2$ are not known with any certainty (Hooper, 1984). All intermediates formed during the conversion of NH_2OH to NO_2^- are believed to remain bound to the complex enzyme hydroxylamine oxidoreductase. The oxidation of NO_2^- to NO_3^- by *Nitrobacter* is a simple two-electron shift in N oxidation state from $+3$ to $+5$ and involves no intermediates (Schmidt, 1982).

There is abundant evidence that both NO and N_2O are usually included among the products of chemoautotrophic nitrification. More than 50 yr ago, Corbet (1935) observed N_2O formation by cultures of nitrifiers supplied with NH_4^+ or NH_2OH, while Ritchie and Nicholas (1972) first reported reduction of NO_2^- to N_2O by dissimilatory NO_2^- reductase synthesized by *Nitrosomonas europaea* in pure culture. Several years later Bremner and Blackmer (1978) recognized that nitrifying microorganisms were responsible for production of a fraction of the soil-emitted N_2O formerly attributed entirely to denitrification. Likewise, it has long been known that NO is produced by chemoautotrophic nitrifiers in culture, but much more recently recognized that this process serves as a significant source of NO emitted from soil. Studies using acetylene or nitrapyrin [2-chloro-6-(trichloromethyl)-pyridine] to inhibit NH_4^+ oxidation and chlorate to inhibit NO_2^- oxidation have demonstrated that both the N_2O (Aulakh et al., 1984a; Blackmer et al., 1980) and NO (Davidson, 1992a; Tortoso & Hutchinson, 1990) produced during chemoautotrophic nitrification are a direct result of the activity of those organisms responsible for the first step of this process, i.e., oxidation of NH_4^+ to NO_2^-.

Recent evidence suggests that production of N_2O by autotrophic NH_4^+ oxidizers in soil results from a reductive process in which the organisms use NO_2^- as an electron acceptor, especially when O_2 is limiting (Poth & Focht, 1985). This mechanism not only allows the organisms to conserve limited O_2 for the oxidation of NH_4^+ (from which they gain energy for growth and regeneration), but also avoids the potential for accumulation of toxic levels of NO_2^-. Our knowledge of nitrifier physiology is not sufficient to predict whether their in situ production of NO also results from NO_2^- reduction as proposed from studies of cell-free extracts by Hooper (1968) and work with intact cells by Remde and Conrad (1990), or if it represents decomposition of an intermediate in the oxidation pathway from NH_4^+ to NO_2^- as proposed from cell-free extract studies by Hooper and Terry (1979). Some evidence indicates that NO production during nitrification also increases as O_2 availability decreases (Lipschultz et al., 1981), but Anderson and Levine (1986) and Remde and Conrad (1990) observed no dependence on the partial pressure of O_2, and it is generally accepted that N_2O production by nitrifiers is more sensitive to this condition. As a result, the NO/N_2O ratio of nitrification products, normally of the order of 10 to 20 in fully aerobic environments (Tortoso & Hutchinson, 1990), decreases along with O_2 partial pressure.

The amounts of nitrification-induced NO and N_2O evolved from soil are regulated by two separate, but interdependent, sets of controllers—those that establish the overall rate of the nitrification process and those that determine the NO/NO_3^- and N_2O/NO_3^- ratios of nitrification products (Firestone & Davidson, 1989). Chemoautotrophic NH_4^+-oxidizing bacteria are widely distributed in soil and require only CO_2, O_2, and NH_4^+ to proliferate. Carbon dioxide is essentially never absent, and O_2 is usually adequate (except for brief intermittent periods) in all but a few very anaerobic environments (e.g., sediments, bogs, sludge), so NH_4^+ availability is the factor that most frequently limits the overall rate of nitrification. Other less important factors that limit nitrifier activity in certain environments include low pH, NO_2^- toxicity, phosphate availability, temperature extremes, low water potential, and allelopathic compounds (Haynes, 1986). The importance of NH_4^+ availability is illustrated in Fig. 5–1, a conceptual model of controls on nitrification reproduced from Robertson (1989). This diagram labels as "proximal" controllers those factors that exert influence at the cellular level and shows pictorially how these are, in turn, regulated by increasingly remote "distal" controllers (mostly soil and environmental parameters). In addition to being essential for all microbial life processes, water is shown to exert an additional important influence on nitrification by controlling the diffusive supply of O_2 and NH_4^+. Similarly, temperature has a modifying influence on both proximal and distal controllers, as well as on the process itself.

Many of the controllers shown in Fig. 5–1 influence not only the overall rate of the nitrification process, but also the NO/NO_3^- and N_2O/NO_3^- ratios of its products. Although control of these ratios is much less well understood, certain relationships are evident. For example, because N_2O apparently results from a reductive process, its importance as a product of nitrification should increase as O_2 availability decreases, but whether that increased importance translates into higher total N_2O production depends on how much the overall process rate is reduced by the limited availability of O_2. The NO and N_2O yields of nitrification are normally relatively small. For N_2O the highest reported value was 20% of the NH_4^+ oxidized in a strongly acid forest soil fertilized with urea (Martikainen, 1985). Such high yields are unusual, and chemical formation of N_2O from NO_2^- may have been involved. The next highest reported yield of N_2O was 6.8% of the urea–N added to a poorly drained Iowa soil (Bremner et al., 1981). The N_2O yield is typically less than 1%, often much less, particularly in well-aerated soil. Because more NO than N_2O is produced under these conditions (Tortoso & Hutchinson, 1990), the NO yield of nitrification varies over a somewhat smaller range. Hutchinson and Brams (1992) and Hutchinson et al. (1992a) found values in the range 1 to 4% of the NH_4^+ oxidized for coarse-textured soils subjected to several N and water treatments, but NO yields as high as 10% (Hutchinson & Follett, 1986; Shepherd et al., 1991) and as low as 0.1% (Davidson et al., 1993) have also been reported.

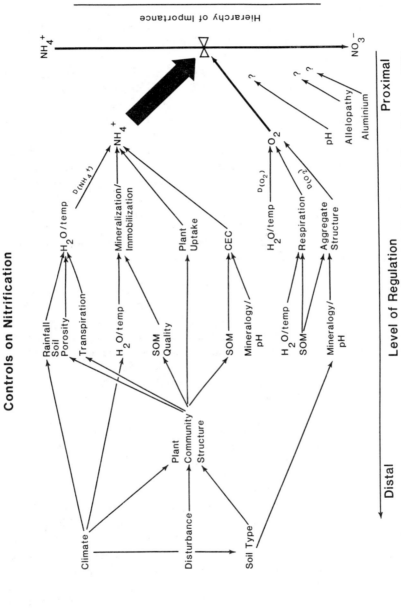

Fig. 5-1. Schematic diagram showing relationships among the major controllers of nitrification in soil (from Robertson, 1989).

Denitrification

For the purposes of this review, denitrification is defined as respiratory reduction of NO_3^- or NO_2^- to gaseous NO, N_2O, or N_2 that is coupled to electron transport phosphorylation. Other anaerobic processes often included in the definition of this term (but excluded here) were described briefly near the beginning of this chapter. Denitrification occupies a position of pivotal importance in the N cycle of the biosphere. In its absence, all biologically available N that has been released from igneous rocks of the earth's original crust and mantle would have been converted long ago to its more thermodynamically stable form of NO_3^- in the oceans (Lindsay et al., 1981). Therefore, this process, which is usually considered to represent loss of biologically available N from terrestrial ecosystems, might alternatively be viewed as the first step toward recovering excess oxidized N by replenishing the supply of atmospheric N_2 available to symbiotic and nonsymbiotic N-fixing microorganisms. Denitrification also represents the only biological process for consumption of N_2O (Firestone & Davidson, 1989), which further enhances its interest to students of global trace N gas dynamics; however, biological uptake and reduction of atmospheric N_2O has been demonstrated outside the laboratory only for wetlands (Ryden, 1981).

Unlike the narrow species diversity of organisms responsible for nitrification in soil, denitrification capacity is common to several taxonomically and physiologically different bacterial groups. Denitrifiers, which are basically aerobic bacteria with the alternative capacity to reduce N oxides when O_2 becomes limiting, are so widely distributed in nature that any restriction of the denitrifying activity in a given habitat can usually be assumed not to result from lack of enzyme, but rather from limited substrate availability or the environmental conditions that regulate the process (Firestone & Davidson, 1989). One possible exception to this generalization is the short-term response to a very large and sudden change in resource availability, but even this exception is likely to be brief. Denitrifying enzyme activity persists for months, even in very dry soil (Smith & Parsons, 1985), and the activation of these enzymes, as well as their de novo synthesis, begins almost immediately following soil wetting by precipitation or irrigation (Rudaz et al., 1991). Hence, when abrupt changes in soil water content create conditions that favor denitrification, its rate is unlikely to be enzyme limited for more than a few hours.

Included among the denitrifiers are phototrophs, lithotrophs, and organotrophs that derive energy for growth and regeneration from light, inorganic substrates, and organic substrates, respectively. The latter group dominates the denitrifying populations of natural soil and water environments. Within this group species of *Pseudomonas* predominate, probably because of their versatility and competitiveness for C substrates, and except in special or unusual environments most of the remaining denitrifying organisms are species of the closely related *Alcaligenes* (Tiedje, 1988). The resulting similarity in the characteristics of numerically dominant populations across a wide range of natural habitats suggests that the controllers of NO and N_2O production (or consumption) by denitrification should also be similar across major global environments (Firestone & Davidson, 1989).

Although the identity of N compounds involved in the biochemical pathway of denitrification is well established, the nature of NO's relation to the process, as well as the exact mechanism for formation of the N–N bond during reduction of NO_2^- to N_2O, are the subjects of considerable current controversy that is beyond the scope of this paper (see Heiss et al., 1989; Weeg-Aerssens et al., 1988; Zafiriou et al., 1989). For our purposes, it is sufficient to know that N_2O is an obligatory intermediate, and that NO behaves as if it were an intermediate, or at least in rapid equilibrium with an intermediate in the reductive sequence (Averill & Tiedje, 1982).

Having already established the near omnipresence of denitrifying enzyme, remaining requirements for this process to occur are availability of suitable reductant (usually organic C), restricted O_2 availability, and presence of N oxides (NO_3^-, NO_2^-, NO, or N_2O). The relative importances of these three denitrification controllers varies among habitats, but for soil and other habitats exposed to the atmosphere, O_2 availability is nearly always the most critical; Tiedje (1988) discussed this topic in detail. Although controls on denitrification at the cellular level are easy to itemize and visualize, environmental factors that regulate these cellular controllers are numerous, interactive, and difficult to conceptualize and model. To illustrate this point, Firestone and Davidson (1989) used the example that in nonflooded terrestrial ecosystems, plant roots can: (i) reduce O_2 availability through root respiration and stimulation of microbial respiration by root exudates, (ii) increase O_2 availability by enhancing gas diffusion rates through removal of soil water, (iii) increase C availability by root exudation, and (iv) decrease N oxide availability through NO_3^- uptake. Robertson's (1989) ordering of the factors regulating denitrification into proximal and increasingly distal classes with a hierarchy of importance established within each class (Fig. 5–2) is probably the most useful and informative conceptual model available for understanding and potentially developing simulation models that encompass the many complex interactions that are possible. Figure 5–2 emphasizes the importance of O_2 availability and shows pictorially how this factor is, for example, strongly influenced by the diffusional constraints imposed by soil water, which is in turn regulated by precipitation, infiltration, and evapotranspiration, which are in turn governed by climate, soil type, plant community structure, etc.

As was the case for nitrification, the total amounts of NO and N_2O produced by denitrification in soil depend not only on factors that determine the overall rate of the process, but also on parameters that control the ratios of its potential products (Firestone & Davidson, 1989). Dinitrogen and N_2O are considered to be the usual end products of denitrification, and that fraction of the process interrupted at N_2O ranges from almost none to the preponderance of N reduced, depending on such factors as N oxide concentration, O_2 availability, organic C availability, soil gas diffusion rates, pH, temperature, sulfide concentration, enzyme activity ratios, and time since initiation of denitrification activity (Tiedje, 1988 and references therein). The

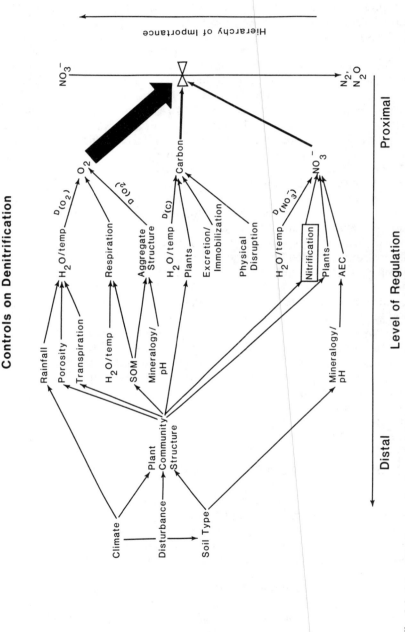

Fig. 5-2. Schematic diagram showing relationships among the major controllers of denitrification in soil (from Robertson, 1989).

regulation imposed by most of these parameters can be explained by considering their influence on the relative availability of oxidant vs. reductant. When the availability of oxidant overshadows the supply of reductant, then substrate N oxide may be incompletely reduced resulting in a larger N_2O/N_2 ratio of end products. Conversely, when the overall rate of denitrification is limited by the supply of oxidant, most of the N oxide is converted to N_2. An analysis of the denitrification sequence based on the Michaelis-Menton model of enzyme kinetics predicts that the proportion of N_2O should increase whenever any of the other controllers listed above slows the overall rate of reduction below the maximum that can be supported by existing enzyme (Betlach & Tiedje, 1981).

Nitric oxide is not usually considered to be a major end product of denitrification in soil or water. Although this belief may partially reflect that convenient and sensitive methodology for NO analysis has been widely available for only the last decade, mass balance experiments long ago confirmed the absence of a major unknown denitrification product. Nevertheless, Tortoso et al. (1986) reported that NO was the principal denitrification product when they initiated the process in laboratory-incubated soil (100 kPa water suction) by removing O_2 from the air stream sweeping the incubation jar headspace. Johansson and Galbally (1984) also found NO to be a major product of denitrification in soil columns incubated under anaerobic flow conditions, and Zafiriou et al. (1989) demonstrated that this gas was the dominant product of denitrification by *P. perfectomarina* when grown in low density suspensions that were highly sparged to remove gaseous products. The condition common to these three data sets that separates them from the body of denitrification literature is that the process was initiated without restricting rapid equilibration of gaseous intermediate products with the ambient atmosphere. In the natural soil environment denitrification generally occurs only when the soil's water content is high enough to restrict O_2 availability, which also restricts the diffusion rates of other gases in soil. The resulting increase in time required for NO diffusion to the soil surface, combined with its instability toward further reduction, allows very little of this gas to escape.

PROCESS STUDIES AT FIELD AND LANDSCAPE SCALES

Despite our increasing understanding of cellular level controls on the microbial production and consumption of gaseous N oxides, applying that knowledge to explain and predict variation in soil NO or N_2O exchange within and across landscapes, or even in small plot studies, remains troublesome. For example, we know that NH_4^+ availability is an important controller of nitrification (Fig. 5-1), as is NO_3^- availability for denitrification (Fig. 5-2), but we lack a universally applicable assay for soil N availability. Robertson and Tiedje (1984) found that a net nitrification assay correlated well with N_2O production in intact cores of forest soils from Michigan, while Matson and Vitousek (1987) reported a good correlation between a net mineralization assay and N_2O flux from tropical forest soils, but not from nearby pastures. In NO emission studies, Williams and Fehsenfeld (1991) found that soil NO_3^- concentration was a good predictor of the differences in emission rates of widely varying ecosystem types across the USA, but Hutchinson et al. (1992b) reported that emissions from a grass pasture on sandy soil in humid subtropical southern Texas were much more strongly related to soil NH_4^+ than soil NO_3^- concentration. All these correlations and others (e.g., Aulakh et al., 1984b; Davidson & Swank, 1986; Rolston et al., 1984) tend to be site-specific or study-specific, and no single predictive parameter or suite of parameters has emerged as applicable across all sites and studies.

Failure to find common predictors probably arises from several confounding factors. First, two very different processes are involved, i.e., nitrification and denitrification. Second, other process-limiting factors that interact with N availability (e.g., the supply of O_2 or organic C) may be more important at some sites than others. In addition, the scale chosen for investigation influences the nature of the predictors likely to be found useful. For example, soil NO_3^- concentration was a good predictor of NO emissions when hardwood forests were compared to fertilized corn (*Zea mays* L.) fields, but it did not account for substantial variation within each location (Williams & Fehsenfeld, 1991); at the latter scale, modest topographic gradients or local-scale effects of crop residues may be important contributors to the observed variability. The relation of NO emissions to soil NO_3^- concentration at the larger scale reflects that this ion generally accumulates when N availability in the system exceeds C availability, resulting in a leaky N cycle. The rate of N cycling may actually be higher in fertile forest sites than in tilled agricultural fields, but high C availability in temperate forest soils usually results in a conservative N cycle, little NO_3^- accumulation, and low emission of gaseous N oxides.

Although the major controllers of soil NO and N_2O exchange may be manifest differently as a function of scale, some generalizations are apparent. Whether it is modeled by Michaelis-Menten kinetics at the cellular level (McConnaughey & Bouldin, 1985) or inferred from stratification by soil orders within a vegetation biome (Matson & Vitousek, 1990), N availability clearly influences gaseous N oxide exchange. Similarly, the total amount of N gases (NO, N_2O, N_2) produced during nitrification and denitrification, as well as the ratios of each to the others, are impacted by the partial pressure of O_2, both when it is measured in a cell culture (Anderson & Levine, 1986; Lipschultz et al., 1981) and when aeration of the soil is crudely indexed by gravimetric soil water content (Davidson, 1992b). Of all the factors known to affect soil NO and N_2O exchange rates, these two are probably the most critical. To understand the variation of N oxide emissions within a cropped field, for example, the distributions of plant roots, crop residues, bands of fertilizer, and small-scale topographic features must all be considered, because they contribute to determining the local N and O_2 supplies. To understand variation within a landscape, soil texture and drainage class have proved useful, presumably as surrogates of O_2 availability (Groffman &

Tiedje, 1989). To improve global budgets of NO and N_2O sources and sinks, global databases of soil orders, climate, and vegetation type may prove to be useful integrators of the effects of N and O_2 availabilities, but we have not yet learned how to apply these data.

SUMMARY

Gaseous N oxides have important impacts on the chemistry of the atmosphere, and their exchange across the surface–atmosphere boundary influences the supply of biologically available N to surface inhabitants. Of the biotic and abiotic processes involved in the production and consumption of NO and N_2O in soil, the bacterial processes of nitrification and denitrification clearly dominate. In both cases, the yields of NO and N_2O depend not only on factors that determine the overall process rates, but also on parameters that control their ratios of potential products.

Nitrification is an aerobic process associated primarily with the metabolism of a group of autotrophic microorganisms with narrow species diversity, but wide distribution. Its magnitude is controlled in most environments by the rate of NH_4^+ supply. Although the NO and N_2O yields of the process are normally relatively small (a few percentage points or less), its near omnipresence in soil makes it a potentially important source in the global atmospheric budgets of the two gases. Because the N_2O (and maybe some of the NO) produced by NH_4^+ oxidizers results from the reduction of product NO_2^-, the total N oxide yield increases and NO/N_2O product ratio decreases with declining O_2 partial pressure.

Denitrification is performed by a diverse and also widely distributed group of aerobic bacteria that have the alternative capacity to reduce N oxides when O_2 becomes limiting. Its rate in soil is most often controlled by O_2 supply, but the availability of suitable reductant (usually organic C) and presence of N oxides (NO_3^-, NO_2^-, NO, or N_2O) are also required. Nitric oxide is not usually considered to be a major end product of denitrification in soil, because the process generally occurs only when soil water content is high enough to restrict O_2 diffusion; the concomitant increase in time required for NO diffusion to the soil surface, combined with its instability toward further reduction, allows very little of this gas to escape. The N_2O/N_2 ratio of denitrification products is controlled primarily by the relative availabilities of oxidant and reductant.

Applying the increasing body of knowledge concerning cellular level controls on the microbial production and consumption of gaseous N oxides at field and landscape scales remains troublesome, but recent recognition of the relationships between the most important cellular controllers (N and O_2 availabilities) and their field, landscape, and global scale manifestations reflects the progress made in this area of research. The challenge facing future research is to integrate knowledge of process controls at all scales to better understand soil emission of NO and N_2O, and then if appropriate, to develop control technologies such as alternative soil management practices or improved fertilizer formulations and application techniques.

REFERENCES

Anderson, I.C., and J.S. Levine. 1986. Relative rates of nitric oxide and nitrous oxide production by nitrifiers, denitrifiers, and nitrate respirers. Appl. Environ. Microbiol. 51:938–945.

Aulakh, M.S., D.A. Rennie, and E.A. Paul. 1984a. Acetylene and N-serve effects upon N_2O emissions from NH_4^+ and NO_3^- treated soils under aerobic and anaerobic conditions. Soil Biol. Biochem. 16:351–356.

Aulakh, M.S., D.A. Rennie, and E.A. Paul. 1984b. Gaseous nitrogen losses from soils under zero-till as compared with conventional-till management systems. J. Environ. Qual. 13:130–136.

Averill, B.A., and J.M. Tiedje. 1982. The chemical mechanism of microbial denitrification. FEBS Lett. 138:8–12.

Betlach, M.R., and J.M. Tiedje. 1981. Kinetic explanation for accumulation of nitrite, nitric oxide, and nitrous oxide during bacterial denitrification. Appl. Environ. Microbiol. 42:1074–1084.

Blackmer, A.M., J.M. Bremner, and E.L. Schmidt. 1980. Production of nitrous oxide by ammonia-oxidizing chemoautotrophic microorganisms in soil. Appl. Environ. Microbiol. 40:1060–1066.

Blackmer, A.M., and M.E. Cerrato. 1986. Soil properties affecting formation of nitric oxide by chemical reactions of nitrite. Soil Sci. Soc. Am. J. 50:1215–1218.

Bleakley, B.H., and J.M. Tiedje. 1982. Nitrous oxide production by organisms other than nitrifiers or denitrifiers. Appl. Environ. Microbiol. 44:1342–1348.

Bollag, J.-M., and G. Tung. 1972. Nitrous oxide release by soil fungi. Soil Biol. Biochem. 4:271–276.

Bormann, F.H., and G.E. Likens. 1979. Pattern and process in a forested ecosystem. Springer-Verlag, Berlin.

Bremner, J.M., and A.M. Blackmer. 1978. Nitrous oxide: Emission from soils during nitrification of fertilizer nitrogen. Science (Washington, DC) 199:295–296.

Bremner, J.M., G.A. Breitenbeck, and A.M. Blackmer. 1981. Effect of anhydrous ammonia fertilization on emission of nitrous oxide from soils. J. Environ. Qual. 10:77–80.

Burth, I., and J.C.G. Ottow. 1983. Influence of pH on the production of N_2O and N_2 by different denitrifying bacteria and *Fusarium solani*. *In* R. Hallberg (ed.) Environmental biogeochemistry. Ecol. Bull. 35:207–215.

Castignetti, D., and H.B. Gunner. 1980. Sequential nitrification by an *Alcaligenes* sp. and *Nitrobacter agilis*. Can. J. Microbiol. 26:1114–1119.

Corbet, A.S. 1935. The formation of hyponitrous acid as an intermediate compound in the biochemical oxidation of ammonia to nitrous acid. II. Microbiological oxidation. Biochem. J. 29:1086–1096.

Davidson, E.A. 1992a. Sources of nitric oxide and nitrous oxide following wetting of dry soil. Soil Sci. Soc. Am. J. 56:95–102.

Davidson, E.A. 1992b. Soil water content and the ratio of nitrous oxide to nitric oxide emitted from soil. *In* R.S. Oremland (ed.) The biogeochemistry of global change: Radiatively active trace gases. Chapman & Hall, New York. (In press.)

Davidson, E.A., P.A. Matson, P.M. Vitousek, R. Riley, K. Dunkin, G. García-Mèndez, and J.M. Maass. 1993. Processes regulating soil emissions of NO and N_2O in a seasonally dry tropical forest. Ecology. (In press.)

Davidson, E.A., J.M. Stark, and M.K. Firestone. 1990. Microbial production and consumption of nitrate in an annual grassland. Ecology 71:1968–1975.

Davidson, E.A., and W.T. Swank. 1986. Environmental parameters regulating gaseous-N losses from two forested ecosystems via nitrification and denitrification. Appl. Environ. Microbiol. 52:1287–1292.

Davidson, E.A., P.M. Vitousek, P.A. Matson, R. Riley, G. García-Mèndez, and J.M. Maass. 1991. Soil emissions of nitric oxide in a seasonally dry tropical forest of Mèxico. J. Geophys. Res. 96:15 439–15 445.

Duxbury, J.M., L.A. Harper, and A.R. Mosier. 1993. Contributions of agroecosystems to global climate change. p. 1–18. *In* L.A. Harper et al. (ed.) Agricultural ecosystem effects on trace gases and global climate change. ASA Spec. Publ. no. 55. ASA, CSSA, SSSA, Madison, WI.

Firestone, M.K., and E.A. Davidson. 1989. Microbiological basis of NO and N_2O production and consumption in soil. p. 7–21. *In* M.O. Andreae and D.S. Schimel (ed.) Exchange of trace gases between terrestrial ecosystems and the atmosphere. John Wiley & Sons, Chichester, England.

Groffman, P.M., and J.M. Tiedje. 1989. Denitrification in north temperate forest soils: Relationship between denitrification and environmental parameters at the landscape scale. Soil Biol. Biochem. 21:621–626.

Haynes, R.J. 1986. Nitrification. p. 127–165. *In* R.J. Haynes (ed.) Mineral nitrogen in the plant-soil system. Academic Press, Sydney.

Heiss, B., K. Frunzke, and W.G. Zumft. 1989. Formation of the N–N bond from nitric oxide by a membrane-bound cytochrome *bc* complex of nitrate-respiring (denitrifying) *Pseudomonas stutzeri*. J. Bacteriol. 171:3288–3297.

Hooper, A.B. 1968. A nitrite-reducing enzyme from *Nitrosomonas europaea*. Preliminaty characterization with hydroxylamine as electron donor. Biochim. Biophys. Acta 162:49–65.

Hooper, A.B. 1984. Ammonia oxidation and energy transduction in the nitrifying bacteria. p. 133–167. *In* W.R. Strohl and O.H. Tuovinen (ed.) Microbial chemoautotrophy. Ohio State Univ. Press, Columbus, OH.

Hooper, A.B., and K.R. Terry. 1979. Hydroxylamine oxidoreductase of *Nitrosomonas*: Production of nitric oxide from hydroxylamine. Biochim. Biophys. Acta 571:12–20.

Hutchinson, G.L., and E.A. Brams. 1992. NO versus N_2O emissions from an NH_4^+-amended Bermuda grass pasture. J. Geophys. Res. 97:9889–9896.

Hutchinson, G.L., and R.F. Follett. 1986. Nitric oxide emissions from fallow soil as influenced by tillage treatments. p. 180. *In* Agronomy abstracts. ASA, Madison, WI.

Hutchinson, G.L., W.D. Guenzi, and G.P. Livingston. 1992a. Soil water controls on aerobic soil emission of gaseous N oxides. Soil Biol. Biochem. (In press.)

Hutchinson, G.L., G.P. Livingston, and E.A. Brams. 1992b. Nitric and nitrous oxide evolution from managed subtropical grassland. *In* R.S. Oremland (ed.) The biogeochemistry of global change: Radiatively active trace gases. Chapman & Hall, New York. (In press.)

Johansson, C., and J.E. Galbally. 1984. Production of nitric oxide in loam under aerobic and anaerobic conditions. Appl. Environ. Microbiol. 47:1284–1289.

Johansson, C., and E. Sanhueza. 1988. Emission of NO from savanna soils during rainy season. J. Geophys. Res. 93:14 193–14 198.

Lindsay, W.L., M. Sadiq, and L.K. Porter. 1981. Thermodynamics of inorganic nitrogen transformations. Soil Sci. Soc. Am. J. 45:61–66.

Lipschultz, F., O.C. Zafiriou, S.C. Wofsy, M.B. McElroy, F.W. Valois, and S.W. Watson. 1981. Production of NO and N_2O by soil nitrifying bacteria. Nature (London) 294:641–643.

Logan, J.A. 1983. Nitrogen oxides in the troposphere: Global and regional budgets. J. Geophys. Res. 88:10 785–10 807.

Martikainen, P.J. 1985. Nitrous oxide emission associated with autotrophic ammonium oxidation in acid coniferous forest soil. Appl. Environ. Microbiol. 50:1519–1525.

Matson, P.A., and P.M. Vitousek. 1987. Cross-system comparisons of soil nitrogen transformations and nitrous oxide flux in tropical forest ecosystems. Global Biogeochem. Cycles 1:163–170.

Matson, P.A., and P.M. Vitousek. 1990. Ecosystem approach to a global nitrous oxide budget. BioScience 40:667–672.

McConnaughey, P.K., and D.R. Bouldin. 1985. Transient microsite models of denitrification: I. Model development. Soil Sci. Soc. Am. J. 49:886–891.

McKenney, D.J., C. Lazar, and W.J. Findlay. 1990. Kinetics of the nitrite to nitric oxide reaction in peat. Soil Sci. Soc. Am. J. 54:106–112.

Melillo, J.M. 1981. Nitrogen cycling in deciduous forests. p. 427–442. *In* F.E. Clark and T. Rosswall (ed.) Terrestrial nitrogen cycles: Processes, ecosystem strategies, and management impacts. Ecol. Bull. 33:427–442.

Nelson, D.W. 1982. Gaseous losses of nitrogen other than through denitrification. p. 327–363. *In* F.J. Stevenson (ed.) Nitrogen in agricultural soils. Agron. Monogr. 22. ASA, Madison, WI.

Poth, M., and D.D. Focht. 1985. ^{15}N kinetic analysis of N_2O production of *Nitrosomonas europaea*: An examination of nitrifier denitrification. Appl. Environ. Microbiol. 49:1134–1141.

Remde, A., and R. Conrad. 1990. Production of nitric oxide in *Nitrosomonas europaea* by reduction of nitrite. Arch. Microbiol. 154:187–191.

Ritchie, G.A.F., and D.J.D. Nicholas. 1972. Identification of the sources of nitrous oxide produced by oxidative and reductive processes in *Nitrosomonas europaea*. Biochem. J. 126:1181–1191.

Robertson, G.P. 1989. Nitrification and denitrification in humid tropical ecosystems: Potential controls on nitrogen retention. p. 55–69. *In* J. Procter (ed.) Mineral nutrients in tropical forest and savanna ecosystems. Blackwell Sci. Publ., Oxford.

Robertson, G.P., and J.M. Tiedje. 1984. Denitrification and nitrous oxide production in successional and old-growth Michigan forest. Soil Sci. Soc. Am. J. 48:383–389.

Robertson, G.P., and J.M. Tiedje. 1987. Nitrous oxide sources in aerobic soils: Nitrification, denitrification, and other biological processes. Soil Biol. Biochem. 19:187–193.

Rolston, D.E., P.S.C. Rao, J.M. Davidson, and R.E. Jessup. 1984. Simulation of denitrification losses of nitrate fertilizer applied to uncropped, cropped, and manure-amended field plots. Soil Sci. 137:270–279.

Rudaz, A., E.A. Davidson, and M.K. Firestone. 1991. Production of nitrous oxide immediately after wetting dry soil. FEMS Microbiol. Ecol. 85:117–124.

Ryden, J.C. 1981. N_2O exchange between a grassland soil and the atmosphere. Nature (London) 292:235–237.

Schimel, J.P., M.K. Firestone, and K.S. Killham. 1984. Identification of heterotrophic nitrification in a Sierran forest soil. Appl. Environ. Microbiol. 48:802–806.

Schmidt, E.L. 1982. Nitrification in soil. p. 253–288. *In* F.J. Stevenson (ed.) Nitrogen in agricultural soils. Agron. Monogr. 22. ASA, Madison, WI.

Shepherd, M.F., S. Barzetti, and D.R. Hastie. 1991. The production of atmospheric NO_x and N_2O from a fertilized agricultural soil. Atmos. Environ. 25:1961–1969.

Smith, M.S., and L.L. Parsons. 1985. Persistence of denitrifying enzyme activity in dried soils. Appl. Environ. Microbiol. 49:316–320.

Smith, M.S., and K. Zimmerman. 1981. Nitrous oxide production by non-denitrifying soil nitrate reducers. Soil Sci. Soc. Am. J. 45:865–871.

Soil Science Society of America. 1987. Glossary of soil science terms. SSSA, Madison, WI.

Stroo, H.F., T.M. Klein, and M. Alexander. 1986. Heterotrophic nitrification in an acid forest soil and by an acid tolerant fungus. Appl. Environ. Microbiol. 52:1107–1111.

Tiedje, J.M. 1988. Ecology of denitrification and dissimilatory nitrate reduction to ammonium. p. 179–244. *In* A.J.B. Zehnder (ed.) Biology of anaerobic microorganisms. John Wiley & Sons, Chichester, England.

Tortoso, A.C., and G.L. Hutchinson. 1990. Contributions of autotrophic and heterotrophic nitrifiers to soil NO and N_2O emissions. Appl. Environ. Microbiol. 56:1799–1805.

Tortoso, A.C., G.L. Hutchinson, and W.D. Guenzi. 1986. Nitric and nitrous oxide emissions during nitrification and denitrification in soil. p. 190–191. *In* Agronomy abstracts. ASA, Madison, WI.

Weeg-Aerssens, E., J.M. Tiedje, and B.A. Averill. 1988. Evidence from isotope labeling studies for a sequential mechanism for dissimilatory nitrite reduction. J. Am. Chem. Soc. 110:6851–6856.

Williams, E.J., and F.C. Fehsenfeld. 1991. Measurement of soil nitrogen oxide emissions at three North American ecosystems. J. Geophys. Res. 96:1033–1042.

Williams, E.J., G.L. Hutchinson, and F.C. Fehsenfeld. 1992. NO_x and N_2O emissions from soil. Global Biogeochem. Cycles. (In press.)

Zafiriou, O.C., Q.S. Hanley, and G. Snyder. 1989. Nitric oxide and nitrous oxide production and cycling during dissimilatory nitrite reduction by *Pseudomonas perfectomarina* J. Biol. Chem. 264:5694–5699.

6

Fluxes of Nitrous Oxide and Other Nitrogen Trace Gases from Intensively Managed Landscapes: A Global Perspective

G. Philip Robertson

W.K. Kellogg Biological Station and Department of Crop and Soil Sciences
Michigan State University
Hickory Corners, Michigan

Concentrations of N_2O in the global atmosphere have been rising over the last 20 yr at about 0.8 parts per billion volumetric (ppb_v) or 0.25% yr^{-1} to the 310 ppb_v present in today's atmosphere (Fig. 6–1; Elkins and Rossen, 1989). Ice core data that indicate concentrations of around 280 ppb_v in the preindustrial atmosphere (e.g., Pearman et al., 1986) suggest that this rise reflects a long-term trend (Fig. 6–2; IPCC, 1990). In fact, this 0.8 ppb_v increase represents a 3.5 Tg addition of N_2O-N to the global atmosphere each year, and if one assumes that the present photodissociation rate of N_2O in the stratosphere (11 Tg yr^{-1} N) is typical of preindustrial steady-state loading rates, then the preindustrial loading rate of N_2O-N was 11 Tg yr^{-1}—suggesting that we are today adding another 40% (3.5 Tg N) of N_2O to the atmosphere each year than was the case 100 yr ago (Cicerone, 1987; Robertson et al., 1989).

This loading rate is significant primarily for two reasons. First, N_2O is one of the major greenhouse gases, accounting for 6 to 8% of the present greenhouse forcing rate ascribed to anthropically derived gases (CO_2, CFC's, CH_4, and N_2O; Hansen et al., 1990; IPCC, 1990). On a molar basis N_2O is about 250 times more potent than CO_2 as an absorber of infrared radiation—in part due to its molecular structure but largely due to the fact that it absorbs in a portion of the infrared transmission window that in our present atmosphere is relatively clean (Duxbury et al., 1993). Second, N_2O is the major natural regulator of stratospheric O_3, which effectively controls the earth's ultraviolet-B (UV_B) radiation balance. Through a series of reactions elucidated over the last 20 yr, the oxidation of N_2O to NO via reaction with photolytically produced atomic oxygen $O(^1D)$ in the upper stratosphere (>25 km) produces NO that in turn reacts with O_3 to form

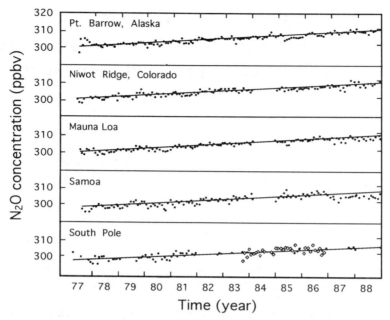

Fig. 6-1. Atmospheric measurements of N_2O from the NOAA-Geophysical Monitoring for Climate Change network (Elkins & Rossen, 1989; redrawn from IPCC, 1990).

NO_2 and O_2 (Hahn & Crutzen, 1981). Under steady-state (preindustrial) conditions, this set of reactions balances the formation of O_3 in the stratosphere. Because these reactions represent the only major atmospheric sink for N_2O, N_2O has an atmospheric lifetime of about 150 yr. This contrasts with a lifetime for CH_4 of about 10 yr, implying that the effects of present N_2O loading rates will be especially long-lasting.

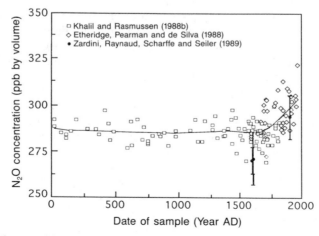

Fig. 6-2. Nitrous oxide concentration in ice core samples (redrawn from IPCC, 1990).

NO_x, on the other hand, is not an effective greenhouse gas at atmospheric concentrations, nor is biogenically produced NO_x an important player in stratospheric gas reactions. Rather, NO_x (primarily NO and NO_2) is important for its role in the chemistry of the troposphere, where it regulates the photochemical production of tropospheric O_3 and the abundance of hydroxl [OH] radicals, primary oxidants for a number of tropospheric trace gases (Jacob & Bakwin, 1990). Under daylight conditions the transformation of tropospheric O_3 to NO_2 and O_2 via reaction with NO is balanced in a 1:1 manner by the photolysis of NO_2 back to NO and O_3; at night there is no reaction of NO_2 back to NO and O_3 so NO_2 may accumulate as O_3 is depleted. Likewise, the reaction of NO with HO_2 in the troposphere is a major regulator of OH concentrations, the primary oxidant for the photochemical destruction of atmospheric CH_4 and other gases. Further, the oxidation of NO_x to HNO_3 in the atmosphere is a major contributor to acid precipitation (NAS, 1986), and may be contributing significantly to low-level N fertilization of large portions of the terrestrial biosphere (Melillo et al., 1989).

Neither the global N_2O budget nor the global NO_x budget is presently balanced with precision—in fact our knowledge of the global N_2O cycle is especially off-track with almost 50% (6.5 Tg N) of the sources needed to balance the known sinks (14.1 Tg N) unidentified (see below). The situation for NO_x is not so bad perhaps only because sinks for NO_x are not as easily quantified as are sinks for N_2O.

In the pages that follow is a discussion of our present state of knowledge with respect to the global budgets of these important trace gases. Of particular note are the roles of agricultural and other intensively managed landscapes as contributors to global fluxes; our knowledge of fluxes in these environments is particularly weak in light of their likely contribution to increased global emission rates.

THE GLOBAL NITROUS OXIDE BALANCE

Of all of the radiatively important trace gases (CO_2, CH_4, CFC, N_2O), none is more poorly understood with respect to sources than N_2O. Unknown sources of N_2O account for almost twice the current global atmospheric loading rate of 3.5 Tg N_2O-N yr^{-1}. Less than 5 yr ago this was not thought to be the case, i.e. the global N_2O budget appeared to be in approximate balance (McElroy & Wofsy, 1986), with 14.1 (± 3.5) Tg yr^{-1} N of sinks about balanced by 15.3 (± 6.7) Tg yr^{-1} of sources (Table 6-1). What has happened since then to shrink our estimates of source–strengths to only 7.5 Tg yr^{-1}? Primarily refinements of existing estimates—and especially refinements of flux estimates from industrial- and managed-landscape sources.

Combustion

Until 1989, global N_2O budgets considered industrial combustion a substantial net source of N_2O to the atmosphere, with an estimated annual con-

Table 6-1. The global N_2O cycle: major sinks and sources (except as noted, references in text).

Sinks/sources	1986[†]	1992
	— Tg yr^{-1} N —	
Sinks (stratosphere)		
Photolysis and 0('D) reactions	10.6	
Atmospheric accumulation (0.25% yr^{-1})	3.5	
Total sinks	14.1	
Sources		
Open ocean	2.0	2.0
Combustion		
Industrial	4.0	0.0
Biomass burning	0.7	0.2 (0.1–0.3)
Manufacturing (nylon)		0.4
Agriculture		
Direct fertilizer emissions	0.8	0.3
Emissions from groundwater		0.2
Indirect via NO_3^- loss		?
Native communities		
Temperate forests, grasslands	0.2–0.6	0.4 (02.–0.6)
Tropical forests	7.4	
Humid forests		2.4
Pasture conversion		0.7
Seasonally dry forests		1.0
Total sources	15.3	7.6
Balance	+1.2	−6.5

† From McElroy and Wofsy (1986).

tribution of approximately 4 Tg N via coal and oil burning. With the discovery (Muzio & Kramlich, 1988; Muzio et al., 1989) that this source estimate is largely a sampling artifact caused by gas reactions in sample flasks prior to N_2O analysis, the strength of this source has been revised downwards almost three orders of magnitude. The present estimate of <0.01 Tg yr^{-1} N_2O–N from industrial combustion reflects this new knowledge (Table 6-1).

Biomass burning—primarily during annual land-clearing operations in tropical savannah and rainforest regions—represents a different combustion source that appears to make a minor though significant contribution to the global N_2O flux. Crutzen and Andreae (1990) estimate a flux of 0.1 to 0.3 Tg yr^{-1} N from tropical sources based on our current knowledge of land clearing rates and N_2O emissions from relatively low-temperature fires. Almost all of this land clearing—certainly most that occurs in extensive savannah regions burned annually or biennially—is associated with agricultural production. In fact, these fires are a very significant source of many other important gases in addition to N_2O; e.g., Crutzen and Andreae (1990) have coined the phrase pyrodenitrification to describe the heretofore unrecognized role of biomass fires in N_2 production.

Row-Crop Agriculture

Row-crop agriculture is assumed in most global assessments of N_2O fluxes to contribute most heavily to the global flux primarily through effects

Table 6-2. Relative effect of different sources of fertilizer on N_2O emissions from fertilized agricultural systems. Control refers to unfertilized crop (from Eichner, 1990).

Treatment (fertilizer type)	Ratio of treatment flux to control flux	Fertilizer-derived N_2O as % of total N_2O emission
Anhydrous ammonia	6.1	82
Ammonium nitrate	2.5	57
Ammonium (Cl, SO_4)	1.4	29
Urea	1.5	30
Nitrate (Ca, K, Na)	1.1	12

of fertilizer inputs. In the mid-1980s approximately 70 Tg of N fertilizer were applied to crops worldwide (FAO, 1985); analysis of the proportion of fertilizer N emitted as N_2O following application (Eichner, 1990) suggests that, on average, about 0.5% (0.35 Tg) of fertilizer inputs is lost from this land use. This estimate is based on evidence that the N_2O flux from fertilized agriculture is most intense following fertilization (references in Eichner, 1990), and is weighted by fertilizer type to account for the fact that N_2O emission rates from fertilizers appear to be strongly related to fertilizer source. For example, direct N_2O fluxes from fields treated with anhydrous ammonia appear to be some six times higher than from fields treated with equivalent amounts of NO_3^- salt (Table 6-2, Fig. 6-3).

Direct fluxes of N_2O from agricultural sources may also result from the hydrologic transport of dissolved N_2O through groundwater to surface waters—where atmospheric equilibration will result in its release to the atmosphere—and to aquifers, where N_2O may be stored as a dissolved gas for long periods (Ronen et al., 1988). The IPCC (1990) global N_2O assess-

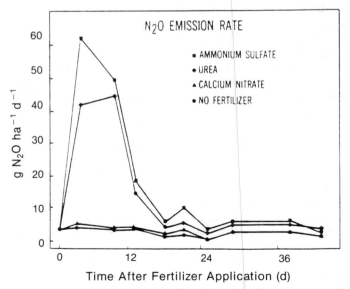

Fig. 6-3. Nitrous oxide emissions from an agricultural field treated with equivalent amounts but different types of N fertilizer (from Breitenbeck et al., 1980).

ment suggested that up to 1.1 Tg N may be added each year to the atmosphere from this source.

The effect of fertilized agriculture on N_2O loss from agricultural landscapes is probably not, however, limited solely to direct losses of N_2O to air and water. Of the more than 70 Tg of fertilizer N added to cropping systems annually, most is not taken off with the crop yield and very little of what remains is retained by the soil–plant system—if it were there would be little need to fertilize annually. Numerous fertilizer-^{15}N studies dating from the mid-1950s (e.g., Allison, 1955; Robertson et al., in review) have demonstrated that less than 50% of the fertilizer N applied to most crops is taken up by the crop. And if one considers that less than 50% of the N that *is* taken up is taken off with the grain yield, and then considers that the N that remains in the residue contributes little to long-term soil organic matter buildup in most tillage systems, then one is left with the conclusion that more than 75% of the N fertilizer added to most cropping systems (50% of that applied + 50% of that taken up by plants) is lost from the residue–soil system on a multiyear basis. At a global level this is a staggering amount of N—approximately 50 Tg yr^{-1}, with most if not all of this N eventually denitrified to close the global N cycle. Only a small portion of this N would need to be denitrified to N_2O rather than to N_2 to have a major impact on the global N_2O budget.

The effect of fertilized agriculture on the global N_2O balance is probably far greater, then, than that projected by current analyses. To evaluate this effect will, however, require analyses of N_2O fluxes at landscape rather than field scales—present studies at field scales, though relatively few, suggest that direct emissions of fertilizer-derived N_2O from agricultural fields are too short-lived to contribute much to annual global losses (e.g., Fig. 6–3). Rather, fertilizer N emissions are likely being expressed primarily at landscape or even regional-level "control points"—habitats such as soil–stream interfaces and near-coastal marine environments where conditions favor the transformation of solute N (NO_3^-, DON, NH_4^+) either directly or indirectly to N_2O and N_2. Too little N_2O flux work has been undertaken along hydrologic flow paths draining agricultural watersheds to adequately evaluate these sources today.

Forests and Grasslands

Estimates of N_2O fluxes from temperate forests and grasslands have changed little since McElroy and Wofsy's (1986) estimates of 0.2 to 0.6 Tg N yr^{-1}. Although it is possible that chronic low-level N fertilization of large regions due to elevated NO_3^- concentrations in rainwater (Melillo et al., 1989) may eventually result in enhanced potentials for N_2O loss from these regions, in the near term it is more likely that the low levels of added N will be immobilized in plant biomass and soil organic matter than be emitted as N_2O (Brumme & Beese, 1992).

Estimates of N_2O fluxes from tropical forests, on the other hand, were probably significantly overestimated in previous budgets. Matson and Vitou-

Table 6-3. Nitrous oxide fluxes from moist and wet lowland tropical forests (from Matson & Vitousek, 1990).

Soil type	Area	N_2O flux	Total flux
	10^6 km^2	kg ha^{-1} yr^{-1}	Tg yr^{-1}
High fertility	1.66	2.6	0.43
Intermediate fertility	9.38	1.9	1.80
Low fertility	1.09	0.5	0.05
Flooded	1.70	0.1	0.02
Other	0.99	0.25	0.03
Total	14.8		2.4

sek (1990) stratified fluxes from intact tropical forest on the basis of soil type, basing an estimate for a given type largely on a generalized relationship between soil N availability (mineralization potential) and N_2O flux for a series of tropical sites (Matson & Vitousek, 1987). This stratification (Table 6-3) leads to a total estimate of 2.4 Tg N_2O-N from moist and wet lowland tropical forests. If one includes Matson and Vitousek's (1990) estimate for additional flux due to conversion of tropical forests to pasture (0.7 Tg N), plus a flux of 1.0 Tg N for fluxes from seasonally dry tropical forests (Vitousek et al., 1989), one can estimate tropical forest fluxes at 4.1 Tg N_2O-N annually. This contrasts with an earlier estimate of 7.4 Tg N for these regions (Table 6-1).

Even this flux may need downward revision, however, when additional measurements of N_2O flux from successional forests become available. Davidson et al. (1991) report higher N_2O fluxes following dry forest conversion to agriculture (*Zea mays*) only for the first 2 yr following clearing; thereafter, following conversion to pasture, N_2O fluxes were similar to the lower fluxes in intact forest. Robertson and Tiedje (1988) found high denitrification rates in lowland tropical rainforests recently cleared of midsuccessional vegetation, but only for the first 6 mo following clearing. After 6 mo and in older midsuccessional sites, rates were substantially lower than in either the recently cut or in uncut primary rainforest sites, apparently due to the rapid depletion of labile soil C stores—initially elevated by clearing—together with the depletion of soil inorganic N due to rapid plant regrowth. Such evidence suggests that tropical forest conversion may lead ultimately to lower rather than to higher N_2O fluxes following deforestation.

Nitrous Oxide

Summing these best estimates of N_2O source strengths, including a newly discovered contribution of 0.4 Tg N from the manufacture of Nylon (Thiemens & Trogler, 1991), yields a total global source strength for N_2O of 7.9 Tg N (Table 6-1). This contrasts with a sink strength of 14.1 Tg N, leaving 6.5 Tg N_2O-N yr^{-1} unidentified as to source.

What is the missing source? Inverse modeling of the global atmosphere using observed interhemispheric N_2O ratios (Prinn et al., 1990) suggests that about one-half of the present atmospheric accumulation rate (3.5 Tg N_2O-

N) ought to be coming from tropical sources and about one-half from temperate. Examination of the global budget presented in Table 6-1 suggests that identified tropical sources are about three times greater than identified temperate sources. This suggests—surprisingly—that our temperate region sources are more poorly known than our tropical sources. Perhaps much of this missing source will be traced to the >70 Tg of manufactured fertilizer N added each year to the biosphere.

THE GLOBAL NITROGEN OXIDES
(NITRIC OXIDE, NITROGEN DIOXIDE) BALANCE

Logan's (1983) global budget for NO_x lists seven sources for atmospheric NO_x that toal to 25 to 99 Tg N yr^{-1} (Table 6-4). Major sinks for NO_x sum to 24 to 64 Tg N. This level of imprecision makes it difficult to evaluate the net global NO_x balance, though one could argue that it is not apparently out of balance.

The large degree of uncertainty in these ranges stem largely from measurement uncertainty, but it also stems from a relatively poor understanding of atmospheric processes affecting NO_x. Unlike N_2O, which is basically unreactive in the lower atmosphere, NO_x reacts with a wide variety of canopy-generated compounds near the ground and is produced elsewhere in the atmosphere by lightning and the oxidation of NH_3. Deposition sinks—in particular dry deposition onto heterogeneous surfaces such as plant canopies—are equally difficult to quantify. The global NO_x budget is correspondingly imprecise.

Of the major biogenic sources of NO_x, soil microbial activity is one of the most important. In a recent revision of Logan's (1983) values for soil NO_x sources, Davidson (1991) revised the earlier estimates of 8 Tg yr^{-1} of emitted NO_x-N to 20 Tg N on the basis of the number of NO_x flux studies

Table 6-4. Nitrogen oxides (NO_x) budget (from Logan, 1983, except as noted).

Sinks/sources	NO_x-N
Sinks	
Wet precipitation	12–42
Dry precipitation	12–22
Total	24–64
Sources	
Combustion	
Industrial	21 (14–28)‡
Biomass burning	12 (4–24)
Lightning	8 (2–20)
Soil (microbial activity)	20†
Atmospheric NH_3 oxidation	6 (1–10)
Oceans	<1
Stratosphere	<1
Total	68 (43–104)

† From Davidson (1991).
‡ Values in parentheses indicate likely range.

published in the interim years. Davidson's analysis suggests that cropland is the most intensive NO_x-N emittor, with average flux rates >6 ng cm^{-2} yr^{-1} that extrapolate to about 30% of the 20 Tg total flux from soils. The 23 flux estimates cited in his analysis represent a major improvement of our understanding of NO_x fluxes from terrestrial ecosystems [a single study was available to Logan (1983)], but the database remains lean; as Davidson (1991) points out, we are in great need of reliable NO_x fluxes from a number of systems and our understanding of the effects of soil and ecosystem disturbance of these fluxes remains weak.

Part of the difficulty in assessing NO_x fluxes from soil stems from the complex reactions that NO_x undergoes in the plant canopy (Jacob & Bakwin, 1991; Johansson, 1989): NO emitted from soils (see Hutchinson & Davidson, 1993) is converted rapidly to NO_2, which can then either be photolyzed back to NO, vented to the atmosphere above the canopy, converted to peroxyacetylnitrate (PAN), or deposited to foliar surfaces where it is reduced to nitrite and then assimilated via nitrite reductase to plant N. The conversion of soil-derived NO to NO_2 is mediated by O_3 and peroxy radicals produced from the decomposition of plant-produced hydrocarbons such as isoprene and terpenes. The conversion of NO_2 to PAN is also mediated by the oxidation of plant-produced hydrocarbons.

The NO_x emissions from terrestrial ecosystems—be they native communities or agricultural communities—is thus a complex function of NO emission rates from soil microbes, plant photosynthetic activity and hydrocarbon emission rates, canopy venting rates, incident light, and tropospheric O_3 concentrations. Understanding the NO_x cycle and its global balance—and especially its capacity for affecting important gas reactions in the troposphere via the regulation of OH and O_3 concentrations—at this point suffers from a paucity of emission data. One hopes that the next 5 yr show as much progress in this area as have the previous 5 yr.

FLUX EXTRAPOLATIONS TO GLOBAL SCALES

A major problem facing those who attempt to use local flux measurements to estimate global fluxes is that of extrapolation. Nitrogen gas fluxes are extremely heterogeneous at both field (e.g., Folonoruso & Rolston, 1984; Robertson et al., 1988) and landscape (e.g., Matson & Vitousek, 1987; Groffman & Tiedje, 1989) scales. Evaluating the appropriate sampling scales for within-field estimation and subsequent extrapolation of these rates to biome/disturbance types worldwide is not an exercise for the faint of heart. It is not unusual for coefficients of spatial variation within individual communities to exceed 100% for chamber-based flux estimates, nor it is unusual for specific types of plant communities (e.g., northern hardwood forest or tropical pasture) to express different annual fluxes in different parts of a landscape.

Problems of extrapolation are best addressed via sampling efforts at a variety of scales, from approaches based on 0.5-m^2 chambers (see Hutch-

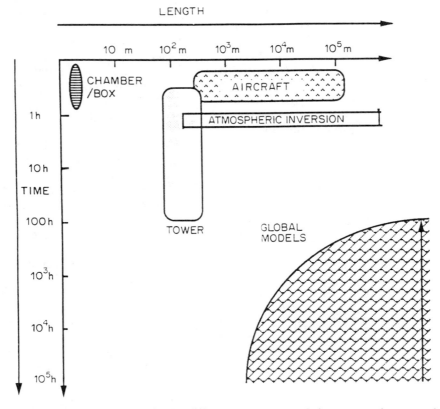

Fig. 6–4. Comparison of the scales that different measurement techniques capture in space and time (from Stewart et al., 1989).

inson & Livingston, 1993) to those employing towers and aircraft with kilometer square and greater footprints (Fig. 6–4; and Desjardin et al., 1993). For N_2O, however, present technology limits sampling strategies primarily to chamber-based approaches: fluxes of N_2O from soils are generally low relative to the 310 ppb_v concentration in today's atmosphere, and this for all practical purposes makes the use of open-air techniques based on concentration gradients (e.g., eddy-correlation and FTIR techniques) all but impossible. We are thus limited to chamber-based approaches for quantifying N_2O fluxes from most ecosystems, which means that our choice of within-system sampling strategies as well as our choice of within-biome habitats will be critical for an adequate assessment of worldwide fluxes—as well as for our predictions of changes in fluxes in response to potentially widespread disturbance such as changes in physical and chemical climate (Robertson et al., 1989).

Up to this point our lack of data from N_2O and NO_x fluxes for major biomes has justified expeditionary-style sampling campaigns to strategic regions. Such campaigns have provided and will continue to provide needed

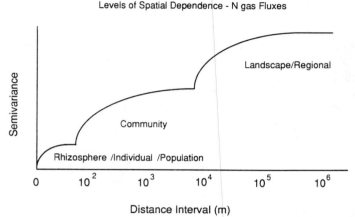

Fig. 6–5. Potential levels of spatial dependence (semivariance) for N gas fluxes across several spatial scales. In this representation >80% of the variance in total global fluxes occurs at the community and landscape/regional scales rather than at the scale of individual plants to populations.

initial assessments of the likely importance of specific biomes and habitat types to regional and global fluxes. Indeed, these assessments form most of the basis for our global flux estimates (Tables 6-1 and 6-4).

We are now close to the point, however, at which much-needed improvements in the precision of our estimates will not come from examination of fluxes in unknown regions, but from refinements of our understanding of fluxes across high-flux landscapes. But these refinements must be sufficiently detailed and must provide sufficient information about mechanisms underlying these fluxes to allow the development of testable gas-flux models. Such models (e.g., Mosier et al., 1983) are necessary for generalizing across both landscapes and disturbance regimes.

Choice of spatial scale is critical for optimizing sampling efforts. One means for making reasonable choices is embedded in geostatistical approaches to assessing spatial variability. In particular, semivariance analysis (e.g., Webster, 1985) provides a means for defining appropriate sample intervals. If, e.g., one were to examine the spatial autocorrelation of a flux across a geographic regions, one would likely find that variance in flux rates changes at definable geographic intervals corresponding to changes in major controls on fluxes.

This can be diagrammed conceptually as in Fig. 6–5, where a stepped semivariance function provides information on the scale at which major controls change (Burrough, 1983). For example, at very small sample intervals control is exerted by rhizosphere or soil aggregate influences as they affect the availability of labile C (provided by root exudates and soil organic matter breakdown), inorganic N supplies, and O_2 availability. Different plant species may have differnet influences on these factors, so that another change in control may appear at scales approximating plant community changes. At greater distances geomorpholocial factors may play a deciding role as soil textures and types change, and at still greater scales climate effects—corresponding to biome changes—will be expressed.

Understanding how variation in fluxes is stepped through this range of scales should give one the opportunity to sample a known proportion of landscape-level variation by sampling at appropriate geographic intervals. For example, landscapes in which 90% of flux variability is expressed at geomorphic scales can be effectively sampled via intense sampling at these larger scales (e.g., Groffman & Tiedje, 1989); landscapes where most flux variability occurs at smaller scales, on the other hand, will require substantially more intensive sampling at the smaller scales.

It is too early to know which fluxes in which landscapes may benefit most from this type of analysis; the formalization of what is now done largely intuitively will probably, however, have the greatest payoff in very heterogeneous landscapes such as those typical of many temperate agricultural regions. Such landscapes are usually made up of a mosaic of very different habitats (field–forest–wetland), with most of the agricultural fields managed very differently from one another with respect to crop rotation and tillage and fertilization regimes.

CONCLUSIONS

1. Global budgets of N_2O and NO_x are important for quantifying the existing impact of anthropic activity on atmospheric processes that affect both physical climate and precipitation chemistry.

2. Current global budgets for N_2O suggest large unknown sources totaling 6.5 Tg N—known sources account for only 7.9 Tg of the 14.1 Tg sink for N_2O-N in the atmosphere.

3. Best estimates of fluxes suggest that managed landscapes—in particular temperate-region landscapes with high fertilizer inputs and tropical landscapes undergoing large-scale land conversion—are the most likely sources behind unknown fluxes.

4. Current global budgets for NO_x are too imprecise to conclusively judge their balance; nevertheless, the impact of soil-produced NO_x likely has a major effect on the OH and O_3 chemistry of the troposphere and on the N content of wet and dry deposition.

5. There appears to be no easy solutions to identifying unknowns in the current N_2O and NO_x budgets. Comprehensive sampling programs must take into consideration large-scale spatial variation (mainly spatial but implicitly temporal) that has been documented for many important habitats.

ACKNOWLEDGMENTS

I thank P.A. Matson and anonymous reviewers for helpful comments on an earlier draft and numerous colleagues for stimulating discussion of the issues raised here. This work was supported by funding from the NSF (LTER Program), the U.S. Dep. of Energy (NIGEC Program), and the Michigan Agricultural Experiment Station.

REFERENCES

Allison, F.E. 1955. The enigma of soil nitrogen balance sheets. Adv. Agron. 7:213–250.

Breitenbeck, G.A., A.M. Blackmer, and J.M. Bremner. 1980. Effects of different nitrogen fertilizers on emission of nitrous oxide from soil. Geophys. Res. Lett. 7:85–88.

Brumme, R., and F. Beese. 1992. Nitrogen fertilization and liming of a temperate forest and release of greenhouse gases. J. Geophys. Res. (In press.)

Burrough, P.A. 1983. Multiscale sources of spatial variation in soil. The application of fractal concepts to nested levels of soil variation. J. Soil Sci. 34:577–597.

Cicerone, R.J. 1987. Changes in stratospheric ozone. Science (Washington, DC) 237:35–41.

Conrad, R., W. Seiler, and G. Bunse. 1983. Factors influencing the loss of fertilizer nitrogen into the atmosphere as N_2O. J. Geophys. Res. 88:6709–6718.

Crutzen, P.J., and M.O. Andreae. 1990. Biomass burning in the tropics: Impact on atmospheric chemistry and biogeochemical cycles. Science (Washington, DC) 250:1669–1678.

Davidson, E.A. 1991. Fluxes of nitrous oxide and nitric oxide from terrestrial ecosystems. p. 219–235. In J.E. Rogers and W.B. Whitman (ed.) Microbial production and consumption of greenhouse gases. Am. Soc. Microbiol., Washington, DC.

Davidson, E.A., P.M. Vitousek, P.A. Matson, R. Riley, G. Garcia-Mendez, and J.M. Maass. 1991. Soil emissions of nitric oxide in a seasonally dry tropical forest of Mexico. J. Geophys. Res. 96(8):15 439–15 445.

Desjardins, R., P. Rochette, I. MacPherson, and E. Patty. 1993. Measurements of greenhouse gas fluxes using aircraft- and tower-based techniques. p. 45–62. In L.A. Harper et al. (ed.) Agricultural ecosystem effects on trace gases and global climate change. ASA Spec. Publ. no. 55. ASA, CSSA, SSSA, Madison, WI.

Duxbury, J.M., L.A. Harper, and A.R. Mosier. 1993. Contributions of agroecosystems to global climate change. p. 1–18. In L.A. Harper et al. (ed.) Agricultural ecosystem effects on trace gases and global climate change. ASA Spec. Publ. no. 55. ASA, CSSA, SSSA, Madison, WI.

Eichner, M.J. 1990. Nitrous oxide emission from fertilized soils: Summary of availabile data. J. Environ. Qual. 19:272–280.

Elkins, J.W., and R. Rossen. 1989. Summary Report 1988: Geophysical monitoring for climatic change. NOAA, ERL, Boulder, CO.

Etheridge, D.M., G.I. Pearman, and F. De Silva. 1988. Atmospheric trace gas variations as revealed by air trapped in an ice core from Law Dome, Antarctica. Ann. Glaciol. 10:38–33.

Food and Agriculture Organization. 1985. Fertilizer yearbook 1984. FAO, Rome.

Folorunso, O.A., and D.E. Rolston. 1985. Spatial variability of field-measured denitrification gas fluxes. Soil Sci. Soc. Am. J. 48:1214–1219.

Groffman, P.M., and J.M. Tiedje. 1989. Denitrification in north temperate forest soils: Relationships between denitrification and environmental factors at the landscape scale. Soil Biol. Biochem. 21:621–626.

Hahn, J., and P.J. Crutzen. 1981. The role of fluxed nitrogen in atmospheric photochemistry. Proc. R. Soc. London, B 196:219–240.

Hansen, J.E., A.C. Lacis, and R.A. Ruedy. 1990. Comparison of solar and other influences on long-term climate. p. 135–145. In K. Schatten and A. Arking (ed.) Climate impact of solar variability. NASA, Washington, DC.

Hutchinson, G.L., and E.A. Davidson. 1993. Processes for production and consumption of gaseous nitrogen oxide in soil. p. 79–93. In L.A. Harper et al. (ed.) Agricultural ecosystem effects on trace gases and global climate change. ASA Spec. Publ. no. 55. ASA, CSSA, SSSA, Madison, WI.

Intergovernmental Panel on Climate Change. 1990. Climate change: The intergovernmental panel on climate change (IPCC) scientific accessment. Univ. Press, Cambridge, England.

Jacob, D.J., and P.S. Bakwin. 1991. Cycling of NO_x in tropical forest canopies. p. 237–254. In J.E. Rogers and W.B. Whitman (ed.) Microbial production and consumption of greenhouse gases. Am. Soc. Microbiol., Washington, DC.

Johansson, C. 1989. Fluxes of NO_x above soil and vegetation. p. 229–246. In M.O. Andreae and D.S. Schimel (ed.) Trace gas exchange between terrestrial ecosystems and the atmosphere. John Wiley, Berlin.

Khahil, M.A.K., and R.A. Rasmussen. 1988. Nitrous oxide: Trends and global mass balance over the last 300 years. Ann. Glaciol. 10:73–79.

Logan, J. 1983. Nitgrogen oxides in the troposphere: Global and regional budgets. J. Geophys. Res. 88:10 785–10 807.

Matson, P.A., and P.M. Vitousek. 1987. Cross-system comparisons of soil nitrogen transformations and nitrous oxide flux in tropical forest ecosystems. Global Biogeochem. Cycles 1:163–170.

Matson, P.A., and P.M. Vitousek. 1990. Ecosystem approach to a global nitrous oxide budget. BioScience 40:677–672.

McElroy, M.B., and S.C. Wofsy. 1986. Tropical forests: Interactions with the atmosphere. p. 33–60. In G.T. Prance (ed.) Tropical rain forests and the world atmosphere. Westview Press, Boulder, CO.

Melillo, J.M., P.A. Steudler, J.D. Aber, and R.D. Bowden. 1989. Atmospheric deposition and nutrient cycling. p. 263–280. In M.O. Andreae and D.S. Schimel (ed.) Trace gas exchange between terrestrial ecosystems and the atmosphere. John Wiley, Berlin.

Mosier, A.R., W.J. Parton, and G.L. Hutchinson. 1983. Modelling nitrous oxide evolution from cropped and native soils. In R. Hallberg (ed.) Environmental biogeochemistry. Ecol. Bull. 35:229–242.

Muzio, L.J., and J.C. Kramlich. 1988. An artifact in the measurement of nitrous oxide from combustion sources. Geophys. Res. Lett. 15:1369–1372.

Muzio, L.J., M.E. Teague, J.C. Kramlich, J.A. Cole, J.M. McCarthy, and R.K. Lyon. 1989. Errors in grab sample measurements of N_2O from combustion sources. J. Air Pollut. Control Assoc. 39:287–293.

National Academy of Sciences. 1986. Acid deposition: Long-term trends. Natl. Acad. Press, Washington, DC.

Pearman, G.I., D. Etheridge, F. DeSilva, and P.J. Fraser. 1986. Evidence of changing concentrations of atmospheric carbon dioxide, nitrous oxide, and methane from air bubbles in antarctic ice. Nature (London) 320:248–250.

Prinn, R., D. Cunnold, R. Rasmussen, P. Simmonds, F. Alyea, A. Crawford, P. Fraser, and R. Rosen. 1990. Atmospheric emissions and trends of nitrous oxide deduced from 10 years of ALE-GAGE data. J. Geophys. Res. 95:18 369–18 835.

Robertson, G.P., M.A. Huston, F.C. Evans, and J.M. Tiedje. 1988. Spatial variability in a successional plant community: Patterns of nitrogen availability. Ecology 69:1517–1524.

Robertson, G.P., and J.M. Tiedje. 1988. Denitrification in a humid tropical rainforest. Nature (London) 336:756–759.

Robertson, G.P., M.O. Andreae, H.G. Bingemer, P.J. Crutzen, R.A. Delmas, J.H. Duyzer, I. Fung, R.C. Harriss, M. Kanakidou, K. Keller, J.M. Melillo, and G.A. Zavarzin. 1989. Trace gas exchange and the physcial and chemical climate: Critical interactions. p. 303–320 In M.O. Andreae and D.S. Schimel (ed.) Trace gas exchange between terrestrial ecosystems and the atmosphere. John Wiley, Berlin.

Robertson, G.P., O.B. Hesterman, and G.H. Harris. Disturbance and synchrony in an agricultural ecosystem. Ecology. (In press.)

Ronen, D., M. Margaritz, and E. Almon. 1988. Contaminated aquifers are a forgotten component of the global N_2O budget. Nature (London) 335:57–59.

Stewart, J.W.B., I. Aselmann, A.F. Bouwman, R.L. Desjardins, B.B. Hicks, P.A. Matson, H. Rodhe, D.S. Schimel, B.H. Svensson, R. Wassmann, M.J. Whiticar, and W.-X. Yang. 1989. Extrapolation of flux measurements to regional and global scales. p. 155–174. In M.O. Andreae and P.S. Schimel (ed.) Exchange of trace gases between terrestrial ecosystems and the atmosphere. John Wiley, Berlin.

Thiemens, M.H., and W.C. Trogler. 1991. Nylon production: An unknown source of atmospheric nitrous oxide. Science (Washington, DC) 251:932–934.

Vitousek, P.M., P.A. Matson, C. Volkman, J.M. Maass, and G. Garcia. 1989. Nitrous oxide flux from dry tropical forests. Global Biogeochem. Cycles 3:375–382.

Webster, R. 1985. Quantitative spatial analysis of soil in the field. Adv. Soil Sci. 3:1–70.

Zardini, D., D. Raynaud, D. Scharffe, and W. Seiler. 1989. N_2O measurements of air extracted from antarctic ice cores: Implications for atmospheric N_2O back to the last glacial-interglacial transition. J. Atmos. Chem. 8:189–201.

7   Cattle Grazing and Oak Trees as Factors Affecting Soil Emissions of Nitric Oxide from an Annual Grassland

Eric A. Davidson

Woods Hole Research Center
Woods Hole, Massachusetts

Donald J. Herman, Ayelet Schuster, and Mary K. Firestone
Department of Soil Science
University of California
Berkeley, California

Range management decisions are driven largely by considerations of animal productivity and land-use regulations, but range management may also impact exchange of trace gases between the biosphere and the atmosphere. Although the dominant source of tropospheric NO in the northern hemisphere is combustion of fossil fuels, soil emissions are not trivial, and NO emissions from grassland soils in rural areas may significantly influence local and regional tropospheric photochemistry of O_3 production and consumption (Logan, 1983). Studies in a variety of ecosystems throughout the world indicate that soil emissions of NO are highly variable and are affected by management regimes (Johansson, 1989).

Cattle (*Bos taurus*) grazing contributed to widespread replacement of native perennial grasses by exotic annual grasses in the central valley of California (Bartolome, 1987). Managers have removed oak (*Quercus* sp.) trees from many oak woodland–grassland ecosystems of California in order to improve grazing productivity (Jackson et al., 1990). That grazing has changed these ecosystems is clear, but the effect of these changes on soil emissions of NO is unknown.

Both cattle and oak trees can be agents of processing and redistribution of N. Grazers consume N in young plant tissue that would otherwise remain sequestered until senescence and decomposition. Nitrogen may accumulate under oak trees as a result of N in litterfall, canopy interception of N deposition, and activity of wild and domestic animals seeking shade. In addition to affecting N availability, oak trees also affect soil moisture and soil temperature that, in turn, may affect soil microbial processes (Callaway et al., 1991; Jackson et al., 1988, 1990). Nitric oxide is emitted by nitrifying and denitrifying soil bacteria and by interactions of biological and abiological soil processes (Firestone & Davidson, 1989).

The first objective of this work is to determine the effects of grazing and vegetation cover on NO emissions at a study site in the central valley of California. The second objective is to investigate the mechanisms that may explain differences in NO fluxes between grazed and ungrazed plots, between plots under oak canopies and plots in open grassy areas, and between wet and dry seasons.

MATERIALS AND METHODS

Site Description

The study site at the University of California Sierra Foothills Range Field Station is located at 200-m elevation in the central valley of California (39°15' N, 121°17' W). The soils in the study watershed are Argonaut silt loams (Mollic Haploxeralfs). A variety of soil properties have been described elsewhere (Davidson et al., 1990; Jackson et al., 1990), including acidity in water of about pH 5.7, and bulk density between 1.3 and 1.4.

The Mediterranean-type climate of this region is characterized by dry hot summers and cool wet winters. Plant communities, described in detail by Jackson et al. (1988, 1990), include blue oak (*Q. douglasii* H. & A.) and live oak (*Q. wislizenii* A.) and a variety of annual grasses and forbs. The species composition and phenology of grasses and forbs differ under oak canopies compared to open grassy areas (Jackson et al., 1988, 1990). These areas will hereafter be called "under-canopy" and "open-grass." Germination of grasses and forbs begins with fall rains, growth continues through the winter, and senescence roughly coincides with cessation of rain in the late spring (Jackson et al., 1988). Ground cover is completely brown during the dry summer. The live oak trees are evergreen; the blue oak trees retain their leaves during the summer and are leafless during the winter.

Sites that were grazed and not grazed by cattle were located within adjacent areas on one of the station's experimental watersheds. Cattle have been rotated into the grazed area for several months each year since 1960. The "ungrazed" area has been fenced and maintained as a natural preserve since 1972. Within each area, two plots (about 2 m by 2 m) were established under blue oak canopies and two in open grass. Four measurements of NO flux were made in each plot in February 1991.

Seasonal variation was studied in another adjacent area that our group has maintained and studied (Davidson et al., 1990). This study site was fenced in 1983 and grazed once in 1984. The site is also instrumented with a datalogger and thermistors at 1-, 4-, and 10-cm depths in both under-canopy and open-grass plots. Six PVC rings (3 under-canopy and 3 open-grass) were installed for NO flux measurements at this site in August 1990 and left in place for repeated measurements in November 1990 and February 1991. At the August date, measurements were repeated four times at each chamber location to investigate diurnal variation. At the other dates and sites, measurements were made as close to midday as possible, and measurements on replicate plots were staggered so as to avoid possible confounding of treatment effects with diurnal soil warming.

Nitric Oxide Flux Measurements

The flux measurement technique has been described in detail by Davidson et al. (1991b). In brief, three or four PVC rings (25-cm diam.) were inserted about 2 cm into the soil in each plot at least 30 min before flux measurements began. A vented plastic cover was placed over the ring (enclosing a volume of about 9 L of air), and air was drawn from the cover (about 0.2 L min^{-1}) by a pump in a chemiluminescence detector (Scintrex LMA-3, Scintrex Limited, Concord, ON, Canada). Air within the cover was mixed with a small fan. The sample air from the cover was mixed with makeup air and passed through a CrO_3 converter that converts NO to NO_2. The air stream was then drawn into the detector where the chemiluminescent reaction of NO_2 with a luminol solution was detected. The NO flux was calculated from the linear increase in the concentration of NO in the air drawn from the cover during the first 5 to 7 min following placement of the cover (Davidson et al., 1991b).

Gross Mineralization and Gross Nitrification by Pool Dilution

Rates of gross mineralization and gross nitrification were measured in the grazed and ungrazed plots in February 1991, when the grasses were in the midst of their growing season. The ^{15}N isotopic pool dilution technique has been described in detail (Davidson et al., 1991a). In brief, solutions of either ($^{15}NH_4$)$_2SO_4$ or $K^{15}NO_3$ (99% enriched, 30 mg N L^{-1}) were injected into intact soil cores (4-cm diam. by 9-cm depth) by making six injections of 1 mL per injection using a spinal needle. One core from each plot was immediately broken up, mixed, and subsampled for extraction in KCl (about 20 g moist soil in 100 mL of 2 M KCl) to determine initial recovery efficiency of added ^{15}N. A second core from each plot was allowed to incubate in the hole from which it came for 24 h. They were then broken up, mixed, and subsampled for KCl extraction.

All KCl extracts were analyzed for NH_4^+ and NO_3^- concentration, using a Lachat flow injection autoanalyzer (Lachat Instruments, Milwaukee, WI), with Cd reduction of NO_3^-. Subsamples of extracts were diffused to concentrate ^{15}N onto filter papers (Brooks et al., 1989) and the ^{15}N enrichment of filter papers was determined by a direct combustion mass spectrometer (Europa Scientific, LTD., Crewe, England).

Rates of gross mineralization were estimated from dilution of the enrichment of the NH_4^+ pool, and gross nitrification was estimated from dilution of the enrichment of the NO_3^- pool. The initial NH_4^+ and NO_3^- pool sizes were estimated from soil sampled from a concentric circle around each soil core. The pool dilution equations of Kirkham and Bartholomew (1954) were used.

Statistical Analyses

The NO flux data were log transformed for statistical analyses. The General Linear Models (GLM) procedure of SAS (SAS Institute, 1985) was used to perform repeated measures tests on diurnal and seasonal variation of NO fluxes. Measurements were repeated at each time on the same three chambers within each vegetation type. The effects of time, vegetation (oak canopy or open grass), and their interaction were tested. For the data collected in February, GLM was used for a two-way ANOVA of grazing and vegetation effects and their interaction. Flux measurements were grouped for each treatment combination, which yielded a sample size of eight for each grazing-by-vegetation treatment.

RESULTS AND DISCUSSION

Diurnal Effects

We measured diurnal variation of NO fluxes in August, when the soil water content was lowest (Table 7–1) and the amplitude of diurnal variation in surface soil temperature was large (Fig. 7–1). Emissions of NO from the open grass peaked during the midafternoon when surface soil temperature also peaked (Fig. 7–2). In contrast, no clear diurnal trend in NO fluxes was observed under an oak canopy. Midafternoon soil temperatures were lower under the oak canopy and the diel amplitude of soil temperature was also smaller (Fig. 7–1). These differences in temperature maxima and in diel amplitude may account for greater diurnal variation and higher NO fluxes in the open-grass area compared to the under-canopy area in August. Both biological and abiological processes of NO production could be temperature dependent, and temperature has been shown to affect diel variation in NO emissions in a variety of ecosystems (Williams & Fehsenfeld, 1991).

The source of high NO emissions from the open-grass plots in August is unknown. The gravimetric soil water content (0.02 kg kg^{-1}, Table 7–1) was below the detection limit of a dew point psychrometer (i.e., < -9 MPa), so microbial processes such as nitrification should have been limited by water stress. Accumulation of soil NO_3^- to concentrations greater than 30 mg N kg^{-1} (Table 7–1), however, indicates that some nitrification must have occurred in the dry soil. The rates of NO_3^- production may be very low in dry soil, but NO_3^- can accumulate in the absence of NO_3^- consumption. One explanation for high NO fluxes may be that the ratio of NO–NO_3^- produced by nitrifying bacteria increases when the bacteria experience water stress.

Table 7–1. Seasonal variation in soil water content and inorganic N concentrations. Data for August and November are from the site fenced since 1983, February data are for the ungrazed site.

Month	Water content		NO_3^-		NH_4^+	
	Canopy	Open	Canopy	Open	Canopy	Open
	— kg H_2O kg^{-1} dry soil		— mg N kg^{-1} dry soil			
August	0.03 (0.00, 3)†	0.02 (0.00, 3)	36.3 (26.7, 3)	43.3 (8.9, 3)	18.5 (0.9, 3)	15.7 (2.1, 3)
November	0.19	0.17	30.6	20.2	5.6	2.4
February	0.35 (0.02, 16)	0.26 (0.01, 8)	2.7 (2.4, 16)	0.4 (0.2, 8)	14.8 (4.7, 16)	5.5 (0.9, 8)

† Means are given, followed by standard errors and the number of replicate samples in parentheses, for August and February data.

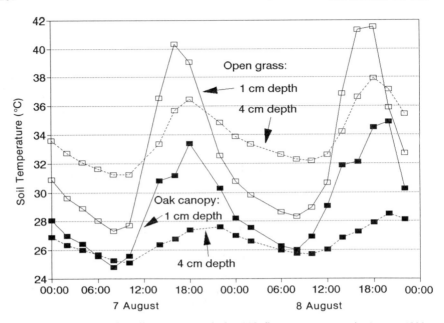

Fig. 7-1. Diel variation in soil temperature during NO flux measurements in August 1990.

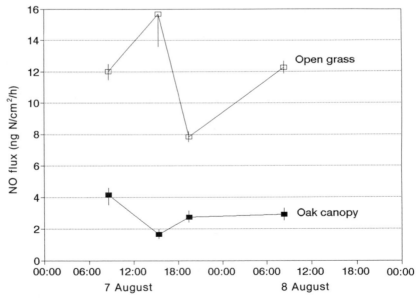

Fig. 7-2. Diurnal variation in NO fluxes in August 1990. Symbols are means and bars are standard errors of three replicates per vegetation type and time. The vegetation effect, the time effect, and the vegetation-by-time interaction effect are significant at $P = 0.01$, 0.05, and 0.10, respectively, by repeated measures ANOVA.

A second explanation is that a significant proportion of the NO_2^- produced by NH_4^+ oxidizing bacteria may be chemodenitrified to produce NO in dry soils (Davidson, 1992). Abiological production of NO (chemodenitrification) resulting from concentration of NO_2^- and H^+ in thin water films of dry soil has been hypothesized (Firestone & Davidson, 1989).

A third explanation also involves chemodenitrification of NO_2^-, but the NO_2^- could be produced by reduction of NO_3^- carried out by thermophilic bacteria. Keeney et al. (1979) showed that chemical decomposition of NO_2^- to NO was very important in sterile soil samples incubated at 40°C and 60°C, and they also hypothesized that biological reduction of NO_3^- to NO_2^- by thermophilic bacteria was followed by abiological decomposition of NO_2^- in nonsterile soils at elevated temperatures. We observed the highest NO emissions when the soil temperature at 1-cm depth was above 40°C in the open grass site (Fig. 7–1), and the temperature at the soil surface was probably higher still.

Diel variation in soil temperature is much less pronounced during the winter and the difference between soil temperature under the oak canopy and in the open grass is small (cf., Fig. 7–1 and 7–3). Soils are wetter in the winter (Table 7–1), and hence have a higher specific heat. In addition, the blue oak canopy is leafless in February.

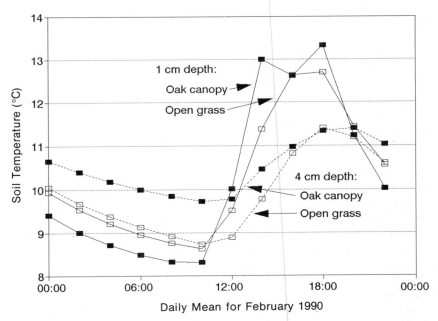

Fig. 7–3. Diel variation in soil temperature in February 1990. The data for each time point are means of soil temperatures recorded during the 28 days of the month. Data from the specific NO sampling date in February 1991, are missing.

Table 7-2. Nitric oxide fluxes, gross rates of N transformations, and inorganic N concentrations measured in February 1991. Gross mineralization, gross nitrification, NH_4^+, and NO_3^- are for the top 9 cm of mineral soil.

Management	Cover type	NO flux	Gross minerali- zation	Gross nitrifi- cation	NH_4^+	NO_3^-
		ng N cm^{-2} h^{-1}	mg N kg^{-1} d^{-1}		— mg N kg^{-1} —	
Grazed	Oak canopy	1.68 (0.36)†‡	24.1 (7.2)	8.2 (4.3)§	9.3 (1.3)	19.0 (12.8)
Grazed	Open grass	1.05 (0.21)	9.7 (3.9)	1.9 (0.0)	6.4 (1.6)	1.1 (0.1)
Ungrazed	Oak canopy	0.75 (0.37)	8.7 (3.6)	2.6 (0.9)	14.8 (1.8)	2.7 (1.5)
Ungrazed	Open grass	0.29 (0.08)	3.1 (0.7)	2.0 (0.2)	5.5 (1.7)	0.4 (0.4)

† Means (and standard errors) are given. Except where otherwise noted, $n = 8$ for NO fluxes, $n = 4$ for other parameters measured in under-canopy plots, and $n = 2$ for other parameters measured in open-grass plots.
‡ $n = 7$.
§ $n = 2$.

Grazing and Canopy Effects—Growing Season

Comparison of grazed and ungrazed plots was made in February, which is the peak of the growing season for grasses in this ecosystem. Soil emissions of NO were higher in grazed than ungrazed plots and higher in plots under oak canopies than those in open grass (Table 7-2). A two-way ANOVA shows that the grazing effect is statistically significant at $P = 0.01$, the canopy effect is significant at $P = 0.06$, and the interaction effect is not significant at $P > 0.10$.

Rates of gross N mineralization were ranked in the same order (highest–lowest) as NO fluxes (Table 7-2). The availability of N, as indicated by gross rates of N mineralization, appears to be an important factor controlling NO fluxes. These data are consistent with the hypothesis that cattle and oak trees increase rates of N cycling. In a larger study encompassing more sampling dates at the same site, Schuster et al. (1991) have found that gross rates of N mineralization and nitrification are consistently higher under oak canopies and within grazed areas. Oak trees may increase N inputs to the soil under the canopy by root uptake of N from beyond the canopy and subsequent deposition of N under the canopy as litterfall. Canopy interception of N in dry deposition is also likely. Cattle shorten the N cycle by digesting grass tissue before it would otherwise senescence and decompose, and then excreting the N in readily available forms with low C–N ratios. If the cattle excrete a disproportionate amount of that N under oak canopies as they seek shade, then oak trees and cattle grazing may have additive effects on N cycling in soil under the canopies.

The grazed canopy treatment also had the highest gross nitrification rate, the highest soil NO_3^- concentration, as well as the highest NO flux and gross N mineralization rate (Table 7-2). In a laboratory study of soils taken from the same watershed, NO production at a variety of soil water contents was shown to be dependent on nitrification (Davidson, 1992). The NO was produced either directly by nitrifying bacteria or by a combination of NO_2^- production by nitrifiers and subsequent chemical decomposition of NO_2^-.

Our data are consistent with the findings of Williams and Fehsenfeld (1991) that show high NO fluxes from soils that accumulate NO_3^- across a broad range of ecosystem types and moisture regimes within the USA. Relatively high soil NO_3^- concentration, as observed under oak canopies in the grazed area, indicates that inorganic N availability exceeds plant and microbial demand for N. Lower belowground inputs of C by rangeland plants and hence lower microbial immobilization of N have been noted as a result of grazing (Holland & Detling, 1990). When inorganic N availability exceeds microbial demand, not only would nitrifying bacteria obtain adequate NH_4^+ substrate to produce NO, NO_2^-, and NO_3^-, but microbial immobilization of much of the NO_3^- produced also would be lower, and NO_3^- may accumulate in the soil. Hence, NO production and NO_3^- accumulation are products of the same mechanistic processes.

Seasonal Effects

The presence of oak canopies had both positive and negative effects on NO fluxes, depending on the season (Fig. 7-4). When temperature differences between open-grass and under-canopy plots were minor in the winter, higher N availability and higher rates of N cycling under oak canopies (Table 7-2) resulted in higher NO fluxes. In contrast, the larger temperature difference between under-canopy and open-grass plots in August may have been a more important factor than N availability in causing higher NO fluxes in the exposed open-grass plots.

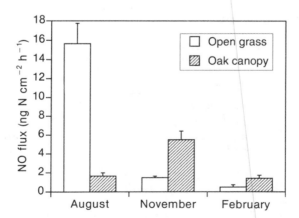

Fig. 7-4. Seasonal differences in vegetation cover effects on NO fluxes. Bars are means and lines are standard errors of three replicate flux measurements per date and vegetation type. The season effect and the season-by-vegetation interaction effect are significant at $P = 0.01$ by repeated measures ANOVA.

Significance of Emissions

Estimates of annual emissions based on infrequent sampling must be interpreted with caution. Emissions of NO during the first rains of the wet season can be important relative to annual estimates (Davidson et al., 1991b), but they were not measured at this site. Despite these limitations, preliminary annual estimates help put the results into appropriate perspective.

Daily means for open grass and under canopy plots (within the area fenced since 1983) in August were assumed to represent the 4-month rainless season (June–September). The November and February sampling dates were assumed to represent four autumn–winter months (October–January) and four spring months (February–May), respectively. Annual emission estimates sum to 0.4 and 0.3 g NO-N m^{-2} for open grass and oak canopy sites, respectively. Assuming 60% oak canopy cover (Jackson et al., 1988), the total soil emissions for the year (in the site fenced in 1983) were about 3 kg NO-N ha^{-1}. This N loss is small relative to internal transformations of N within the ecosystem (Davidson et al., 1990), but is significant relative to watershed estimates of 6.9 kg N ha^{-1} yr^{-1} input via atmospheric deposition and 5 kg N ha^{-1} yr^{-1} lost in streams (Dahlgren & Singer, 1991).

Grazing has increased availability of inorganic N, and hence has increased at least one N-loss mechanism. Soil emissions of NO were about twice as high in grazed plots as in ungrazed plots during the growing season of grasses in February, although we do not know if this trend persists throughout the year. Moreover, the long-term effects of these higher soil emissions of NO on total N and site fertility are not known.

The importance of an increased soil source of NO depends on the relative importance of the soil source compared to fossil fuel sources. In California, fossil fuel sources dominate. If these results are applicable to other grazed grassland systems, then the additional NO flux to the atmosphere caused by grazing in many rural regions could significantly affect local tropospheric processes of O_3 production.

ACKNOWLEDGMENTS

The authors thank L. Cox, M. Coyne, K. Dunkin, T. Holden, D. Hooper, and M. Kaufman for help with field work. Eric Davidson thanks the National Research Council Associateships Program for support during the period of field work. This work was supported by NSF grant BSR-88-08187 and IHRMP project 89009.

REFERENCES

Bartolome, J.W. 1987. California annual grassland and oak savannah. Rangelands 9:122–125.

Brooks, P.D., J.M. Stark, B.B. McInteer, and T. Preston. 1989. Diffusion method to prepare soil extracts for automated nitrogen-15 analysis. Soil Sci. Soc. Am. J. 53:1707–1711.

Callaway, R.M., N.M. Nadkarni, and B.E. Mahall. 1991. Facilitation and interference of *Quercus douglasii* on understory productivity in central California. Ecology 72:1484–1499.

Dahlgren, R., and M.J. Singer. 1991. Nutrient cycling in managed and unmanaged oakwoodland-grass ecosystems. p. 337–341. *In* R.B. Stundiford (ed.) Proc. of Symp. on Oak Woodlands and Hardwood Rangeland Management, Davis, CA. 31 Oct.–2 Nov. 1990. Gen. Tech. Rep. PSW-126 USDA-FS, Pacific Southwest Res. Stn., Berkeley, CA.

Davidson, E.A. 1992. Sources of nitric oxide and nitrous oxide following wetting of dry soil. Soil Sci. Soc. Am. J. 56:95–102.

Davidson, E.A., S.C. Hart, C.A. Shanks, and M.K. Firestone. 1991a. Measuring gross nitrogen mineralization, immobilization, and nitrification by ^{15}N isotopic pool dilution in intact soil cores. J. Soil Sci. 42:335–349.

Davidson, E.A., J.M. Stark, and M.K. Firestone. 1990. Microbial production and consumption of nitrate in an annual grassland. Ecology 71:1968–1975.

Davidson, E.A., P.M. Vitousek, P.A. Matson, R. Riley, G. Garcia-Mendez, and J.M. Maass. 1991b. Soil Emissions of nitric oxide in a seasonally dry tropical forest of Mexico. J. Geophys. Res. 96:15 439–15 445.

Firestone, M.K., and E.A. Davidson. 1989. Microbiological basis of NO and N_2O production and consumption in soil. p., 7–21. *In* M.O. Andreae and D.S. Schimel (ed.) Exchange of trace gases between terrestrial ecosystems and the atmosphere. John Wiley & Sons, New York.

Holland, E.A., and J.K. Detling. 1990. Plant response to herbivory and belowground nitrogen cycling. Ecology 71:1040–1049.

Jackson, L.E., R.B. Strauss, M.K. Firestone, and J.W. Bartolome. 1988. Plant and soil nitrogen dynamics in a California annual grassland. Plant Soil 110:9–17.

Jackson, L.E., R.B. Strauss, M.K. Firestone, and J.W. Bartolome. 1990. Influence of tree canopies on grassland productivity and nitrogen dynamics in deciduous oak savanna. Agric. Ecosyst. Environ. 32:89–105.

Johansson, C. 1989. Fluxes of NO_x above soil and vegetation. p. 229–246. *In* M.O. Andreae and D.S. Schimel (ed.) Exchange of trace gases between terrestrial ecosystems and the atmosphere. John Wiley & Sons, New York.

Keeney, D.R., I.R. Fillery, and G.P. Marx. 1979. Effect of temperature on the gaseous nitrogen products of denitrification in a silt loam soil. Soil Sci. Soc. Am. J. 43:1124–1128.

Kirkham, D., and W.V. Bartholomew. 1954. Equations for following nutrient transformation in soil, utilizing tracer data. Soil Science Sci. Soc. Am. Proc. 18:33–34.

Logan, J. 1983. Nitrogen oxides in the troposphere: Global and regional budgets. J. Geophys. Res. 88:10785–10807.

SAS Institute. 1985. SAS user's guide: Statistics. Version 5 ed. SAS Institute, Cary, NC.

Schuster, A., K.A. Dunkin, J.M. Stark, E.A. Davidson, D. Herman, and M.K. Firestone. 1991. Cows and trees accelerate rates of nitrogen cycling in a California oak woodland-grass ecosystem. Bull. Ecol. Soc. Am. 72:244.

Williams, E.J., and F.C. Fehsenfeld. 1991. Measurement of soil nitrogen oxide emissions at three North American ecosystems. J. Geophys. Res. 96:1033–1042.

8   Denitrification in Subsurface Environments: Potential Source for Atmospheric Nitrous Oxide

Charles W. Rice and Kirby L. Rogers

Department of Agronomy
Kansas State University
Manhattan, Kansas

Atmospheric concentrations of radiatively active trace gases, such as nitrous oxide (N_2O) are increasing because of more sources, fewer sinks, or both. Unaccounted N_2O sources need to be identified to explain these increasing atmospheric concentrations (Davidson, 1991). Until recently, biological activity was considered negligible below the active root zone. Recent studies have discovered a diverse microbial ecosystem to depths of 200 m (Balkwill & Ghiorse, 1985; Fliermans, 1989; Wilson et al., 1990). Denitrifiers are important components of this environment. Francis et al. (1989) measured denitrifying activity in sediments from >250 m to bedrock (Fig. 8–1). Denitrifying activity was measured throughout the profile except in the drier clay formations. Thus, denitrification below the active root zone in the subsurface environment is one possible source of atmospheric N_2O. Aquifers combined with volcanoes and lightning have been estimated to contribute approximately 0.8 Tg N yr^{-1}, which represents 5 to 10% of the total global N_2O source (Davidson, 1991; Wuebbles & Edmonds, 1991). The contribution of subsoils and aquifers to atmospheric N_2O may be underestimated and increasing because of contamination of subsurface environments. Ronen et al. (1988) estimated that 10 to 20% of biogenic N_2O sources could originate in aquifers.

For the purpose of this discussion, the subsurface environment will be divided into two zones. The first zone includes the subsoil and the underlying intermediate vadose zone, which is the unsaturated layer between the active root zone and the saturated layer. Most biological measurements are made in the active root zone, generally in the top 60 to 90 cm of the soil profile. Measurement of N_2O flux from the soil surface may account for part of the N_2O produced below the active root zone, but the contributing factors affecting N_2O production will be different in subsoils than in the surface layer of soil. The second zone is the groundwater or aquifer system. Lowrance and Pionke (1989) have provided a detailed discussion of the aquifer system

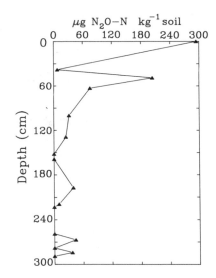

Fig. 8-1. Profile distribution of denitrification activity under anaerobic conditions of deep sediments collected from a site in the southeastern coastal plains (from Francis et al., 1989).

in relation to N transformations. Previous reviews have discussed denitrification in subsoils and the groundwater as a mechanism for NO_3^- removal; we will focus on denitrification as a potential source of atmospheric N_2O.

SUBSOILS AND THE INTERMEDIATE VADOSE ZONE

Disappearance of NO_3^- and decreasing NO_3^-/Cl^- ratios are the most common indirect indicators of subsoil denitrification. This assumption may be reasonable below the active root zone, where plant uptake is not a major mechanism of NO_3^- removal. Several studies using NO_3^-/Cl^- ratios have demonstrated subsoil denitrification in some soils (Gast et al., 1974; Gambrell et al., 1975; Devitt et al., 1976). Gambrell et al. (1975) found very little evidence for denitrification in a well-drained subsoil; but in a poorly drained soil, denitrification was prevalent in a zone with a seasonal water table. Isotopic fractionation of NO_3^- also can be used to indicate previous denitrification activity. Denitrification will result in an enrichment of the heavier isotope, ^{15}N. Gormly and Spalding (1979) demonstrated denitrification as a major mechanism of loss in the intermediate vadose zone using isotopic fractionation. Although isotopic fractionation of NO_3^- and NO_3^-/Cl^- ratios may be good indirect indicators of previous denitrification in the profile, these techniques cannot detect the gaseous end products, which are important for determining global N_2O budgets.

More direct evidence for denitrifying activity is provided by measurement of denitrifying enzymes or incubation of samples to measure the nitrogenous gas products. McGarity and Myers (1968) demonstrated for denitrification by the latter method (Fig. 8-2). Significant denitrifying activity was stratified and variable within the soil profile, but the increased

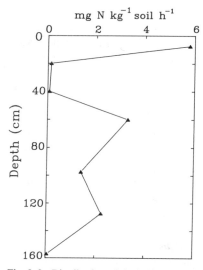

Fig. 8-2. Distribution of denitrification activity in a soil profile with high levels of subsoil denitrification (from McGarity & Myers, 1968).

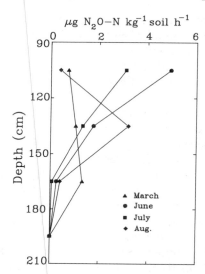

Fig. 8-3. Distribution of denitrifying enzyme activity for the Kahola silt loam soil profile.

activity shown in Fig. 8-2 was associated with the beginning of the B_{2t} horizon. Significant subsoil denitrifying activity was attributed to transitory accumulations of organic C in the subsoil by leaching of highly available soluble organic C from the overlying surface horizons (Myers & McGarity, 1971). In a Kahola silt loam soil (fine-silty, mixed, mesic Cumulic Hapludolls), we also have found denitrifying enzyme activity to be stratified within the soil profile (Fig. 8-3). The stratification was associated with physical changes, such as bulk density and increased clay in the subsoil (Rice & Rogers, 1991). Not all subsoils demonstrate significant denitrification. Unstratified coarse-textured soils generally do not exhibit significant denitrification apparently because of the lack of organic C or anaerobic conditions (Lund et al., 1974; Devitt et al., 1976; Parkin & Meisinger, 1989; Rice & Rogers, 1991). Fine-textured soils also may not have significant subsoil denitrification apparently because of the lack of NO_3^- or organic C moving into the subsoil (Gast et al., 1974).

Denitrification below the active rooting zone appears to be controlled by several factors. Biological denitrification requires the presence of: (i) NO_3^-; (ii) appropriate bacteria; (iii) an energy source, usually organic C; and (iv) anaerobiosis. The lack of denitrifying bacteria does not appear to be a major limitation based on previous results (Parkin & Meisinger, 1989; Rice & Rogers, 1991). Both heterotrophic and autotrophic denitrifiers exist, but heterotrophs, such as *Pseudomonas,* are usually more abundant. Temperatures of 15 °C, commonly encountered in the intermediate vadose zone, will decrease the rate, but not inhibit denitrification. Based on observations of subsoil denitrification, one of the most important variables creating conditions conducive for denitrification is stratification of morphological properties within the soil profile. Changes in soil physical properties create discontinuities that impede water and solute movement. Layers that impede water movement allow the soil above the discontinuity to have higher soil moisture contents, resulting in decreased O_2 diffusion and increased potential for anaerobic microsites. The morphological characteristics that result in these discontinuities include textural changes (Lund et al., 1974; Devitt et al., 1976) and bulk density. Lund et al. (1974) reported a good relationship between the average clay content and NO_3^- concentration in the control section (1.8- to 8-m depth) of the soil profile. Significantly lower NO_3^- concentrations than expected based on the clay content were found in two soils that had variable textures within the control section or contained a duripan. These two situations impede water movement, thus enhancing subsoil denitrification. Devitt et al. (1976) similarly concluded that subsoil denitrification was possible in coarse-textured soils that contained a subsurface layer of higher clay content. Saturation does not always indicate conditions favorable for denitrification. During high-intensity precipitation events, denitrification was apparently repressed by transport of dissolved O_2 with the infiltrating water (Gambrell et al., 1975). Redox measurements are often an indicator of reducing conditions. A redox of < 350 mV is often used as the point for denitrification; however, a lower redox reading does not mean denitrification necessarily will occur, unless the other requirements for denitrification, i.e., NO_3^-, appropriate bacteria, and energy source, also are present.

In addition to anaerobic conditions, soluble organic C and NO_3^- can accumulate at the zones of restricted water movement. McGarity and Myers (1968) reported a good correlation between total organic C and denitrification in the surface horizons but a lack of fit in subsoils. Myers and McGarity (1971) later concluded that the transitory nature of denitrifying activity in the subsoil was supported by transport of soluble organic C from the organically rich surface horizons. The accumulation of dissolved organic C at zones of restricted water movement also may create anaerobic microsites within an aerobic soil (Rice et al., 1988). Organic C and saturation were the primary factors determining the presence of denitrification in the zone of a seasonal water table (Gambrell et al., 1975).

Lind (1983) suggested that chemodenitrification may occur in subsoils. Buresch and Moraghan (1976) demonstrated chemodenitrification of NO_3^- by ferrous iron (Fe^{2+}) only in the presence of Cu^{2+}. They concluded that for most subsoils, however, NO_3^- reduction was not likely due to inadequate concentrations of Fe^{2+} and Cu^{2+}. The importance of chemodenitrification deep in soil profiles may need further research. The potential for reduction of NO_2^- by Fe^{2+} is greater, but the gas product is pH dependent and also may require a catalyst such as Cu^{2+} (Nelson & Bremner, 1970; Moraghan & Buresch, 1977). Chemodenitrification of NO_3^- for the production of N_2O probably is not considered a major process. The role of autotrophic denitrifiers associated with Fe chemistry is not known.

The amount of subsoil denitrification that will result in N_2O production is not known. The ratio of N_2O/N_2 is variable because it is affected by NO_3^- concentration, diffusion, and other factors (Firestone, 1982). Most studies have focused on total denitrification, because the primary focus has been the removal of NO_3^- to maintain groundwater quality. The indirect indicators for denitrification, NO_3^- disappearance, NO_3^-/Cl^- ratios, and redox potentials, do not measure the gaseous products. Lind and Eiland (1989) measured N_2O production to a depth of 20 m with samples that were incubated anaerobically (Fig. 8–1). The addition of a C source enhanced N_2O production in the subsurface environment, which indicates the potential for N_2O production. Reports of N_2O concentrations in soil profiles in situ are rare. Rolston et al. (1976) measured N_2O concentrations of a repacked Yolo loam (fine-silty, mixed, nonacid, thermic Typic Xerorthents). Maximum concentrations up to 32 μg N_2O-N mL^{-1} soil air were reported at 85 cm. In the field, N_2O concentrations of 9.6 μg N_2O-N mL^{-1} soil air were greatest near the soil surface and declined with depth. The authors also indicated that the peak N_2 concentration occurred 4 d after the peak N_2O concentration. These results suggest the potential for N_2O to be removed from the site of denitrification by diffusion or leaching of dissolved N_2O before it is further reduced to N_2. Rice and Rogers (1991) measured N_2O concentrations in an undisturbed soil profile of a Kahola silt loam. The N_2O profile demonstrates: (i) a significant amount of gaseous N_2O throughout the soil profile, and (ii) an increase in N_2O at certain zones related to stratified soil layers (Fig. 8–4). Because N_2O is soluble in water, the dissolved N_2O is susceptible to downward movement. Bowden and Bormann (1986) measured a 10- to 100-fold increase in N_2O concentrations up to 300 μg N_2O-N mL^{-1} in the soil water after tree harvest in a forested watershed.

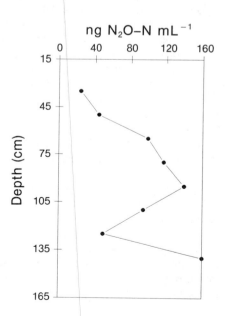

Fig. 8–4. Nitrous oxide concentrations in the soil atmosphere of a Kahola silt loam soil profile.

GROUNDWATER

To determine the production of N_2O in groundwater, the geology, hydrology, and geochemistry of the groundwater must be considered. Recent reviews of denitrification of groundwater have been published in the context of NO_3^- contamination (Lowrance & Pionke, 1989; Hiscock et al., 1991). Consideration of these subject areas will be briefly discussed in this paper, but more detailed reviews should be consulted.

In a confined aquifer containing high NO_3^- levels of natural origin, research by Vogel et al. (1981) suggested very slow rates of denitrification. Evidence for denitrification was based on dissolved O_2 and N_2 in the aquifer and the isotopic fractionation ($^{15}N/^{14}N$ ratio) of the dissolved N_2 gas. Initiation of denitrification occurred only after 13 000 yr, when the O_2 supply was nearly depleted. In a confined Lincolnshire aquifer, Wilson et al. (1990) attributed the decrease in NO_3^- concentrations to denitrification based on increases in dissolved N_2 and light $\delta^{15}N$ values of the N_2 gas. Conflicting evidence for denitrification in the Chalk aquifer in England has been reported (Foster et al., 1985; Howard, 1985). Based on geochemistry and presence of denitrifying bacteria, Foster et al. (1985) suggested that denitrification was responsible for removal of NO_3^- from recharge water entering the Chalk aquifer. Based on NO_3^- disappearance, the in situ rate of NO_3^- reduction was calculated at 1.5 mg N m^{-2} d^{-1}, which corresponds to 5.5 kg N ha^{-1} yr^{-1}. Howard (1985) concluded that denitrification was not a major process after assessing the age and hydrogeochemistry of the waters in the Chalk aquifer, however. A more complete analysis of several case studies of different aquifers is presented by Lowrance and Pionke (1989). Denitrification in confined aquifers appears to be variable and may not produce significant quantities of N_2O. Any N_2O that is produced is not likely to interact readily with the atmosphere, unless through outgassing at wells. In confined aquifers, N_2O is more likely to be reduced to N_2 because it would have greater residence time than in subsoils or unconfined aquifers.

Shallow unconfined aquifers are more likely sources of N_2O, because they are likely to have greater C levels, to receive NO_3^--contaminated water from anthropogenic sources, and to interface with the atmosphere. Trudell et al. (1986) conducted an excellent study in which they injected NO_3^- and Br$^-$ into a nearly static, shallow, unconfined, sand aquifer. By measuring the disappearance of NO_3^- compared to the conservative tracer, Br$^-$, they were able to determine denitrification rates. These ranged up to 0.13 g N m^{-3} h^{-1}, which corresponds to 3.9 kg N ha^{-1} h^{-1} for the entire 4-m deep aquifer. More direct assays for denitrification involve measuring denitrification under anaerobic conditions. Adelman et al. (1986) measured denitrification rates of 0.05 to 1.46 mg N L^{-1} d^{-1} in samples incubated anaerobically from an anoxic, shallow, sandy aquifer. Slater and Capone (1987) reported a denitrification rate of 238 μg N_2O–N kg^{-1} d^{-1} from samples taken at a 4-m depth into an aquifer. Denitrification was NO_3^--limited rather than C-limited in this particular aquifer. Shallow aquifers of riparian zones associated with agricultural watersheds also may be sources of N_2O, because they often receive high concentrations of NO_3^- and usually have an abundant source of C. Jacobs and Gilliam (1985) determined that the majority of the 10 to 55 kg NO_3^-–N ha^{-1} yr^{-1} lost from an agricultural field by subsurface drainage was removed by denitrification in riparian zones. Peterjohn and Correll (1984) estimated that 67% of NO_3^- removed in a riparian zone was due to denitrification. Groffman et al. (1991) reported a low potential for denitrification in a forested vegetative filter strip, however. A fluctuating water table also can be a potential site for enhanced denitrification. The interface between the saturated and unsaturated layers was a prime site for denitrification in a poorly drained soil in North Carolina (Gambrell et al., 1975).

The other likely source for subsurface N_2O production is contaminated subsoils and aquifers. Ronen et al. (1988) estimated that 3.4 to 7.8 kg N_2O–N ha^{-1} yr^{-1} was produced in contaminated aquifers and may be an overlooked source for increasing atmospheric N_2O. An aquifer that received sewage effluent containing high NO_3^- concentrations had high denitrifying activity that removed up to 1 mM NO_3^- d^{-1} (Smith & Duff, 1988). Denitrification was primarily C-limited because of an excess amount of NO_3^- entering the aquifer from the sewage effluent. In some aquifers, C could be the primary contaminant, thus resulting in NO_3^--limiting conditions, however. Addition of NO_3^- has been proposed as a means to remediate aquifers contaminated with organics (Widdowson et al., 1988; Kinzelbach et al., 1991).

Other possible mechanisms of NO_3^- reduction prevent the exclusive use of NO_3^- disappearance and NO_3^-/Cl^- ratios to indicate denitrification. The reduction of NO_3^- to NH_4^+ through dissimilatory NO_3^- reduction is an alternative fate of NO_3^- (Simmons et al., 1985). This does not appear to be a major process in most aquifers where the C-rich and highly anaerobic conditions (Tiedje, 1988) that favor it are not likely to occur. Dissimilatory NO_3^- reduction to NH_4^+ would not contribute to global N_2O. Chemical denitrification as a result of Fe^{2+} oxidation also is possible in some aquifers, where pyrite is present (Postma et al., 1991). Apparently, incomplete oxidation of pyrite by O_2 increases Fe^{2+}, which then can react with NO_3^-. The gaseous end products from this reaction still need further research, as discussed earlier. Postma et al. (1991) indicated that, in most aquifers, the reduction of NO_3^- by reduced metal cations is rarely significant because of the lack of available metals to reduce contaminant levels of NO_3^-.

The important regulatory factors for N_2O production in aquifers need to be defined to assess the impact of geology and anthropogenic activities on denitrification and N_2O production. Recent studies indicate that microbial populations (Balkwill & Ghiorse, 1985; Fliermans, 1989; Wilson et al., 1990) and denitrifiers (Foster et al., 1985; Francis et al., 1989; Obenhuber & Lowrance, 1991) are present and active in many aquifers. How ubiquitous denitrifiers are in aquifers is not known. Organic C often is the most limiting factor for microbial activity in aquifers. Spalding et al. (1978) reported that significant amounts of organic C from decaying alfalfa (*Medicago sativa* L.) roots and manure applications may have been transported from the overlying soil through the vadose to the aquifer. Apparently, soluble organic C either was not interacting with or was bypassing, through macroporous flow, the intermediate vadose zone. Those authors indicated that 2 mg C L^{-1} was sufficient to support denitrification in aquifers. In a study by Francis et al. (1989), sediments to depths of 200 m were primarily NO_3^--limited rather than C-limited. The greatest response was to NO_3^- additions in all samples at all depths. Stimulated denitrifying activity from additions of C occurred only in samples from unsaturated zones in near-surface formations. Thurman (1985) suggested that the source of organic C in deep subsurface material was degradation of fossilized organic matter in aquifers. The texture of the geologic material within the aquifer and fluctuations in water table height also may promote denitrification by increasing the potential for anaerobiosis (Gambrell et al., 1975; Lowrance & Pionke, 1989).

Dissolved N_2O concentrations in aquifers have been measured to a limited extent. Obenhuber and Lowrance (1991) measured 1.22 μg N_2O-N L^{-1} with a range of 0 to 8.7 μg N_2O-N L^{-1} in an aquifer in the Georgia Coastal Plain. The concentration of N_2O increased with depth within the aquifer, suggesting that N_2O production occurred within the aquifer and dissolved N_2O was not transported from the overlying soil. In aquifer microcosms, production of N_2O was enhanced by the addition of glucose. Wilson et al. (1990) also measured elevated levels of N_2 and N_2O in some aquifers, which were attributed to denitrification in water containing high amounts of NO_3^-. The elevated levels were greater than would be expected at normal atmospheric pressures and water temperatures. Ronen et al. (1988) estimated that N_2O concentrations dissolved in the groundwater were three orders of magnitude greater than if equilibrated with the atmosphere. Enhanced N_2O concentrations were attributed to anthropogenic activities at the soil surface, such as waste application, cultivation, and fertilization.

FATE OF NITROUS OXIDE

To have an impact on atmospheric concentrations, N_2O produced in subsurface environments obviously must be released to the atmosphere. Nitrous oxide may diffuse upward to the soil surface, or N_2O dissolved in the water may be transported along hydrologic gradients (Fig. 8–5). Diffused N_2O can be measured when it evolves from the soil surface; however, its source cannot be identified, especially if underlying subsoils or groundwaters are contaminated and producing significant quantities of N_2O. Dissolved N_2O in the water also may be transported from the site of production, which complicates measurements (Burke & Lashof, 1990). In experiments with a Histosol, leachate contained 11 to 52% of the total N_2O produced in drained columns and 78 to 98% in flooded columns (Guthrie & Duxbury, 1978). During transport through the intermediate vadose zone or along the hydrologic gradient of the aquifer, N_2O may be further reduced to N_2.

The N_2O in the aquifer may diffuse upward into the vadose zone and potentially to the soil surface. Dissolved N_2O in the aquifer also may be

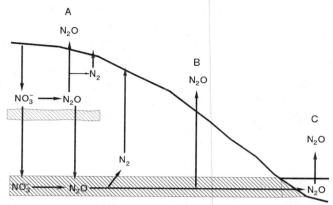

Fig. 8–5. Source, transport, and fate of N_2O in the subsurface environment. Sources of N_2O to the atmosphere include diffusion to the surface from (A) stratified soil horizons, (B) aquifers, and (C) outgassing from aquifers in contact with surface waters.

eventually released to the atmosphere through outgassing at interfaces with surface waters or from wells (Bowden & Bormann, 1986; Ronen et al., 1988). Slater and Capone (1987) found measurable denitrification at the point of submarine discharge of NO_3^--contaminated groundwater. It is also likely that N_2O could outgas at such discharge locations. Drainage water from organic soils in New York contained 200 to 900 mL N_2O L^{-1}, which was 900 to 4200 times greater than expected for water in equilibrium with the atmosphere. Seeps from harvested forest watersheds contained dissolved N_2O two orders of magnitude greater than expected, and the N_2O rapidly degassed 10 m from the seep (Bowden & Bormann, 1986). They suggested that outgassing of transported dissolved N_2O in shallow aquifers may be a major source for atmospheric N_2O. Locating and quantifying these sources of N_2O release may be important, because the gas may be transported from the site of production.

RESEARCH NEEDS

Research is needed to determine the extent of N_2O production in subsurface environments to better estimate its potential contribution to atmospheric N_2O. Assessing this contribution involves problems in quantification beyond those normally encountered for surface environments. These problems are accentuated by the difficulty of obtaining samples to represent spatial and temporal variability. Denitrification appears to be the major source of N_2O in the subsurface, but the ratio of N_2O/N_2 needs to be determined to correlate denitrification to N_2O production. The regulatory factors for N_2O production in the subsurface environment, such as physical discontinuities that affect the distribution and transmission of O_2, NO_3^-, and organic C, need to be identified. Although denitrifiers have been reported in many subsurface environments, their ubiquity needs to be confirmed. The impact of anthropogenic activities on N_2O production in subsurface ecosystems needs to be assessed, because NO_3^- contamination is increasing. Soil management practices, land application of wastes, and fertilization will impact and potentially increase N_2O production. The magnitude of this effect is currently not known. As interest increases to promote denitrification to remediate NO_3^-- (Hiscock et al., 1991; Obenhuber & Lowrance, 1991) and organic-contaminated groundwaters (Kinzebach et al., 1991), the completeness of the NO_3^- reduction process needs to be evaluated to determine the effect on atmospheric N_2O.

SUMMARY

Subsurface environments, including the intermediate vadose zone and aquifers, may be contributing to increased atmospheric concentrations of N_2O. Denitrification appears to be the major source of N_2O in the subsurface environment. In the intermediate vadose zone, the level of denitrifying activity is dependent on the soil morphology, particularly stratified layers within the soil profile, which impede water and solute movement and create conditions favorable for denitrification. Movement of organic C from the soil surface appears to support denitrifying activity by providing an energy source and increasing the consumption of O_2. Denitrification and N_2O production have been observed in aquifers but appear to be of greatest significance in shallow unconfined aquifers. The lack of organic C, NO_3^-, or anaerobiosis is often a limiting factor for activity but seems to be site specific. The presence of denitrifying bacteria does not appear to be a major limitation, based on published results, but the ubiquity of denitrifiers in subsurface environments needs to be confirmed. The fate of the N_2O produced in subsurface environments is unknown. Transport of N_2O by upward diffusion, by outgassing at contacts with surface waters, and by groundwater use need to be quantified to determine the contribution to atmospheric N_2O. Contamination of subsurface environments with NO_3^- and organics has the potential for increasing the contribution to atmospheric N_2O by enhancing denitrification.

ACKNOWLEDGMENTS

Support for research was provided by grants from the Kansas Water Resources Research Institute. Contribution no. 92-309-B from the Kansas Agricultural Experiment Station.

REFERENCES

Adelman, D.D., M.D. Jawson, and J.S. Schepers. 1986. Measurement of denitrification in the eastern sandhills region of Nebraska. p. 457–465. *In* Proc. Agricultural Impacts on Ground Water—A conf., Omaha, NE. 11–13 August. Water Well J. Publ. Co., Dublin, OH.

Balkwill, D.L., and W.C. Ghiorse. 1985. Characterization of subsurface bacteria associated with two shallow aquifers in Oklahoma. Appl. Environ. Microbiol. 50:580–588.

Bowden, W.B., and F.H. Bormann. 1986. Transport and loss of nitrous oxide in soil water after forest clear cutting. Science (Washington, DC) 233:867–869.

Buresch, R.J., and J.T. Moraghan. 1976. Chemical reduction of nitrate by ferrous iron. J. Environ. Qual. 5:320–325.

Burke, L.M., and D.A. Lashof. 1990. Greenhouse gas emissions related to agriculture and land-use practices. p. 27–43. In B.A. Kimball et al. (ed.) Impact of carbon dioxide, trace gases, and climate change on global agriculture. ASA Spec. Publ. 53. ASA, Madison, WI.

Davidson, E.A. 1991. Fluxes of nitrous oxide and nitric oxide from terrestrial ecosystems. p. 219–235. In J.E. Rogers and W.B. Whitman (ed.) Microbial production and consumption of greenhouse gases: Methane, nitrogen oxides, and halomethanes. Am. Soc. Microbiol., Washington, DC.

Devitt, D., J. Letey, L.J. Lund, and J.W. Blair. 1976. Nitrate nitrogen movement through soil as affected by soil profile characteristics. J. Environ. Qual. 5:283–288.

Firestone, M.K. 1982. Biological denitrification. p. 289–326. In F.J. Stevenson (ed.) Nitrogen in agricultural soils. Agron. Monogr. 22. ASA, Madison, WI.

Fliermans, C.B. 1989. Microbial life in the terrestrial subsurface of southeastern coastal plain sediments. Hazard. Waste & Hazard. Mater. 6:155–171.

Foster, S.S.D., D.P. Kelly, and R. James. 1985. The evidence for zones of biodenitrification in British aquifers. p. 356–369. In D.B. Caldwell et al. (ed.) Planetary ecology. Van Nostrand Reinhold Co., New York.

Francis, A.J., J.M. Slater, and C.J. Dodge. 1989. Denitrification in deep subsurface sediments. Geomicrobial J. 7:103–116.

Gambrell, R.P., J.W. Gilliam, and S.B. Weed. 1975. Denitrification in subsoils of the North Carolina coastal plain as affected by soil drainage. J. Environ. Qual. 4:311–316.

Gast, R.G., W.W. Nelson, and J.M. MacGregor. 1974. Nitrate and chloride accumulation and distribution in fertilized tile drained soils. J. Environ. Qual. 3:209–213.

Gormly, J.R., and R.F. Spalding. 1979. Sources and concentrations of nitrate-nitrogen in ground water of the central Platte region, Nebraska. Ground Water 17:291–301.

Groffman, P.M., E.A. Axelrod, J.L. Lemunyon, and W.M. Sullivan. 1991. Denitrification in grass and forest vegetated filter strips. J. Environ. Qual. 20:671–674.

Guthrie, T.F., and J.M. Duxbury. 1978. Nitrogen mineralization and denitrification in organic soils. Soil Sci. Soc. Am. J. 42:908–912.

Hiscock, K.M., J.W. Lloyd, and D.N. Lerner. 1991. Review of natural and artificial denitrification of groundwater. Water Res. 25:1099–1111.

Howard, K.W.F. 1985. Denitrification in a major limestone aquifer. J. Hydrol (Amsterdam) 76:265–280.

Jacobs, T.C., and J.W. Gilliam. 1985. Riparian losses of nitrate from agricultural drainage waters. J. Environ. Qual. 14:472–478.

Kinzelbach, W., W. Schafer, and J. Herzer. 1991. Numerical modeling of natural and enhanced denitrification processes in aquifers. Water Resour. Res. 27:1123–1135.

Lind, A.M. 1983. Nitrate reduction in the subsoil. p. 145–156. In H.L. Golterman (ed.) Denitrification in the nitrogen cycle. Plenum Press, New York.

Lind, A.M., and F. Eiland. 1989. Microbiological characterization and nitrate reductions in subsurface soils. Biol. Fertil. Soils 8:197–203.

Lowrance, R.R., and H.B. Pionke. 1989. Transformations and movement of nitrate in aquifer systems. p. 373–392. In R.F. Follett (ed.) Nitrogen management and ground water protection. Elsevier Sci. Publ. Co., New York.

Lund, L.J., D.C. Adriano, and P.F. Pratt. 1974. Nitrate concentration in deep soil cores related to soil profile characteristics. J. Environ. Qual. 3:78–82.

McGarity, J.W., and R.J.K. Myers. 1968. Denitrifying activity in solodized solonetz soils of eastern Australia. Soil Sci. Soc. Am. Proc. 32:812–817.

Moraghan, J.T., and R.J. Buresch. 1977. Chemical reduction of nitrite and nitrous oxide by ferrous iron. Soil Sci. Soc. Am. J. 41:47–50.

Myers, R.J.K., and J.W. McGarity. 1971. Factors influencing high denitrifying activity in the subsoil of solodized solonetz. Plant Soil 35:145–160.

Nelson, D.W., and J.M. Bremner. 1970. Role of soil minerals and metallic cations in nitrite decomposition and chemodenitrification in soils. Soil Biol. Biochem. 2:1–8.

Obenhuber, D.C., and R. Lowrance. 1991. Reduction of nitrate in aquifer microcosms by carbon additions. J. Environ. Qual. 20:255–258.

Parkin, T.B., and J.J. Meisinger. 1989. Denitrification below the crop rooting zone as influenced by surface tillage. J. Environ. Qual. 18:12–16.

Peterjohn, W.T., and D.L. Correll. 1984. Nutrient dynamics in an agricultural watershed: Observations on the role of a riparian forest. Ecology 65:1146–1157.

Postma, D., C. Boesen, H. Kristiansen, and F. Larsen. 1991. Nitrate reduction in an unconfined sandy aquifer: Water chemistry, reduction processes, and geochemical modeling. Water Resour. Res. 27:2027–2045.

Rice, C.W., and K.L. Rogers. 1991. Denitrification in the vadose zone: A mechanism for nitrate removal. Contribution no. 287. Kansas Water Resour. Res. Inst., Kansas State Univ., Manhattan.

Rice, C.W., P.E. Sierzega, J.M. Tiedje, and L.W. Jacobs. 1988. Stimulated denitrification in the microenvironment of a biodegradable organic waste injected into soil. Soil Sci. Soc. Am. J. 52:102–108.

Rolston, D.E., M. Fried, and D.A. Goldhamer. 1976. Denitrification measured directly from nitrogen and nitrous oxide gas fluxes. Soil Sci. Soc. Am. J. 40:259–266.

Ronen, D., M. Magaritz, and E. Almon. 1988. Contaminated aquifers are a forgotten component of the global N_2O budget. Nature (London) 335:57–59.

Slater, J.M., and D.G. Capone. 1987. Denitrification in aquifer soil and nearshore marine sediments influenced by groundwater nitrate. Appl. Environ. Microbiol. 53:1292–1297.

Simmons, J.A.K., T. Jickells, A. Knap, and W.B. Lyons. 1985. Nutrient concentrations in ground waters from Bermuda: Anthropogenic effects. p. 383–398. In D.G. Caldwell et al. (ed.) Planetary ecology. Van Nostrand Reinhold Co., New York.

Smith, R.L., and J.H. Duff. 1988. Denitrification in a sand and gravel aquifer. Appl. Environ. Microbiol. 54:1071–1078.

Spalding, R.F., J.R. Gormly, and K.G. Nash. 1978. Carbon contents and sources in ground water of the central Platte region in Nebraska. J. Environ. Qual. 7:428–434.

Tiedje, J.M. 1988. Ecology of denitrification and dissimilatory nitrate reduction to ammonium. p. 179–243. In A.J. Zehnder (ed.) Biology of anaerobic microorganisms. John Wiley and Sons, New York.

Thurman, E.M. 1985. Organic geochemistry of natural waters. Matinus Nijhoff/Dr. W. Junk, Dordrecht, The Netherlands.

Trudell, M.R., R.W. Gillham, and J.A. Cherry. 1986. An in-situ study of the occurrence and rate of denitrification in a shallow unconfined sand aquifer. J. Hydrol. (Amsterdam) 83:251–268.

Vogel, J.C., A.S. Talma, and T.H.E. Heaton. 1981. Gaseous nitrogen as evidence for denitrification in groundwater. J. Hydrol. (Amsterdam) 50:191–200.

Widdowson, M.A., F.J. Molz, and L.D. Benefield. 1988. A numerical transport model for oxygen- and nitrate-based respiration linked to substrate and nutrient availability in porous media. Water Resour. Res. 24:1553–1565.

Wilson, G.B., J.N. Andrews, and A.H. Bath. 1990. Dissolved gas evidence for denitrification in the Lincolnshire limestone aquifer groundwater, eastern England. J. Hydrol. (Amsterdam) 113:51–60.

Wuebbles, D.J., and J. Edmonds. 1991. Primer on greenhouse gases. Lewis Publ., Chelsea, MI.

9 Nitrous Oxide Emissions and Methane Consumption in Wheat and Corn-Cropped Systems in Northeastern Colorado

K.F. Bronson

International Rice Research Institute
Manila, Philippines

A.R. Mosier

USDA-ARS
Fort Collins, Colorado

Agricultural soils serve as a significant source of atmospheric nitrous oxide (N_2O), and can be sinks for atmospheric methane (CH_4) in aerated soils (Bouwman, 1990). Nitrous oxide emissions during both oxidation of NH_4 and denitrification (Bouwman, 1990) are increased by N fertilizer applications to soil (Eichner, 1990). Nitrous oxide, an important "greenhouse gas," is 300 (mass basis) times more radiatively active than CO_2 (Rodhe, 1990), and participates in ozone (O_3) destruction in the stratosphere (Crutzen, 1976). Globally, N_2O flux from fertilized soils is estimated to be 1 to 2 Tg N_2O–N yr^{-1} (Seiler and Conrad, 1987).

Methane also is a greenhouse gas that is about 15 times more effective than CO_2 (mass basis) at absorbing infrared radiation (Rodhe, 1990). Recent decreases in the CH_4 consumption or sink activity in soils, in addition to sources such as rice (*Oryza sativa* L.) paddies, cattle, landfills, and biomass burning, might account for the 1% yr^{-1} increase in CH_4 in the atmosphere (Ojima et al., 1992). Consumption of atmospheric CH_4 in soil has been reported in a variety of ecosystems (Steudler et al., 1989; Keller et al., 1990; Whalen & Reeburgh, 1990; Mosier et al., 1991). Methane oxidation or consumption in soils is done by a diverse group of methanotrophic bacteria (Whittenbury et al., 1970) and by nitrifying bacteria (Jones & Morita, 1983). Nitrogen fertilizer, especially NH_4-based fertilizers, have recently been reported to strongly repress CH_4 consumption in forest (Steudler et al., 1989) and prairie (Mosier et al., 1991) soils.

 Agricultural Ecosystem Effects on Trace Gases and Global Climate Change. ASA Special Publication no. 55.

In this paper, recent research on N_2O emissions and CH_4 consumption in wheat (*Triticum aestivum* L.) and corn (*Zea mays* L.)-cropped soils of the Northern Plains is summarized. Soil properties, such as texture, organic matter content, and inorganic N concentrations are discussed in terms of how they control the flux rates of these trace gases.

MATERIALS AND METHODS

Dryland Wheat–Fallow Rotation

Six plots were established in a wheat field and in an adjacent fallow field in June 1990 in the Central Plains Experimental Range in northeastern Colorado (Mosier et al., 1991). The soil is a sandy clay loam (fine-loamy, mixed, mesic Ustollic Haplargid) that has not been fertilized. Nitrous oxide and CH_4 fluxes were measured weekly by taking samples from vented chambers (0, 15, and 30 min) after placing them over the soil (Hutchinson & Mosier, 1981). The samples were analyzed by gas chromatography (Mosier and Mack, 1980; Mosier et al., 1991). Cumulative flux for the growing season was calculated for each plot by linearly interpolating data points and integrating the area by Simpson's rule. Soil samples were taken at each gas sampling date from the 0- to 15-cm layer were analyzed for 2 M KCl–extractable NH_4 and NO_3 by a flow-injection autoanalyzer. Soil moisture was determined gravimetrically, and soil temperature was measured at the 2.5-cm depth by digital thermometer on each gas sampling date.

Irrigated Wheat

Nitrous oxide and CH_4 fluxes were measured (1–3 times wk^{-1}) from a flood-irrigated wheat field from planting in September 1990 to harvest in July 1991 from vented chambers with 1-h incubation periods. Cumulative flux was calculated as described above. The clay soil (fine, montmorillonitic, mesic Aridic Arginstoll) at the site had 60 mg NO_3–N kg^{-1} in the surface 30 cm. Four blocks in the experiment received the following fertilizer treatments 1 d after planting: urea alone, urea plus the nitrification inhibitors (NIs) encapsulated Ca carbide (ECC) (Banerjee & Mosier, 1989) at 20 kg CaC_2 ha^{-1} or dicyandiamide (DCD) (10% of N rate), and unfertilized control. The urea-fertilized treatments were applied at the rates of 50, 100, or 150 kg N ha^{-1}. Soil moisture and exchangeable N were determined periodically, as described above, in the 0- to 15- and 15- to 30-cm layers of soil. Soil temperature was measured at the 7.5-cm depth of soil. Additionally, a total denitrification study was conducted adjacent to the experiment after the irrigation on 230 days after fertilization (DAF) using the CaC_2–C_2H_2 inhibition technique (Aulakh et al., 1991) to determine N_2O–(N_2 + N_2O) ratios during denitrification in this soil. Six 180-cm^2 plots received 50 mg NO_3–N, and three of these had CaC_2 added prior to 1-h incubations in vented covers.

Irrigated Corn

Periodic (weekly or less) measurements of N$_2$O and CH$_4$ were made on four replicates of a furrow-irrigated corn field from planting in April 1990 to harvest in September 1990, and cumulative flux calculated as described above (Bronson et al., 1992). The clay loam soil (fine, montmorillonitic, mesic Aridic Argiustoll) at this site had 6 mg NO$_3$–N kg^{-1} in the surface 25 cm. Fertilizer treatments (at time of planting) included 0 and 218 kg urea–N ha^{-1} and applications of the NIs nitrapyrin [2-chloro-6-(trichloromethyl)-pyridine] (0.5 L a.i. ha^{-1}) and ECC (20 kg CaC$_2$ ha^{-1}). Soil water, temperature, and exchangeable N were determined as described in the irrigated wheat.

RESULTS AND DISCUSSION

Dryland Wheat–Fallow Rotation

Nitrous Oxide Emissions

Nitrous oxide emissions in dryland wheat–fallow system were low throughout fall and winter (0–240 d after planting) (Fig. 9–1a), with no differences between wheat-cropped and fallow soil ($P > 0.05$). After two spring rains, large N$_2$O fluxes were measured in both wheat-cropped and fallow soils. Nitrate–nitrogen in the surface 15 cm of soil was apparently not related to N$_2$O fluxes ($P > 0.05$). Relatively high levels of NO$_3$–N were present in the wheat-cropped soil (27 mg NO$_3$–N kg^{-1}), which probably accumulated during mineralization of organic N during the previous fallow. Soil moisture (Fig. 9–1c) was higher in the fallow field ($P < 0.05$) throughout the growing season, but it was not related to N$_2$O emissions ($P > 0.05$).

Methane Consumption

Methane consumption rates were about 1 g CH$_4$–C ha^{-1} for the fall and winter in both wheat-cropped and fallow soils (Fig. 9–1b). The adjacent grassland, of a similar soil type, however, (last cropped 50 yr ago and allowed to go back to native grass species) consumed CH$_4$ at about twice this rate (Ojima et al., 1992). By early summer when soil moisture was declining in the wheat-cropped soil, CH$_4$ consumption increased to a maximum of 4.9 g CH$_4$–C ha^{-1} d^{-1} 302 d after planting. Methane consumption was negatively correlated with soil moisture (Fig. 9–1c) in the 0- to 15-cm layer of soil ($r = -0.55$, $n = 21$, $P < 0.01$) after 259 d after planting, but was not related with NH$_4$ or NO$_3$ in soil ($P > 0.05$) at any time.

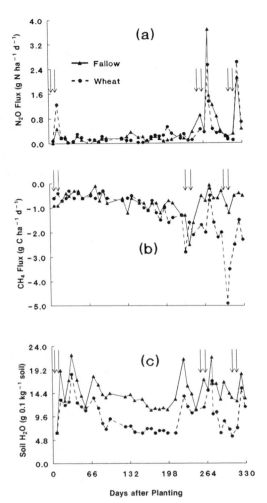

Fig. 9-1. Trace gas fluxes in wheat and fallow cropping systems: (a) N_2O, (b) CH_4, and (c) soil H_2O in surface 2.5 cm. Double arrows indicate a rainfall >25 mm.

Irrigated Wheat

Nitrous Oxide Emissions

Emissions of N_2O were highest early in the growing season and after irrigations (Fig. 9-2a). In the first 60 DAF, fluxes were highest with urea alone, compared to urea plus NIs ($P < 0.05$). Aulakh et al. (1984) and Magalhaes et al. (1984) reported that NIs can indirectly reduce N_2O emissions by retarding NH_4 oxidation in soil. The N_2O emitted in the fall, in the first 60 DAF was probably largely from nitrification of NH_4 (Bremner & Blackmer, 1978). This can be inferred because ample NO_3-N (60 mg kg^{-1}) (Fig. 9-2c)

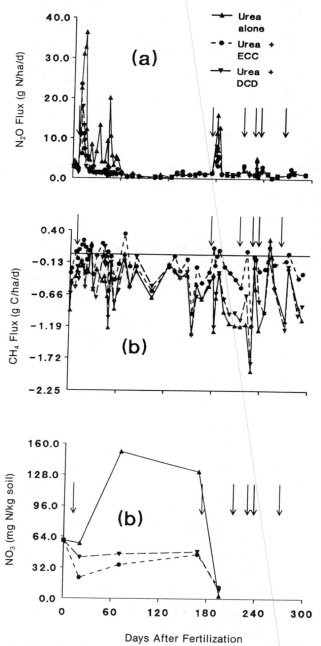

Fig. 9–2. Trace gas fluxes from irrigated wheat as affected by urea with and without encapsulated calcium carbide (ECC) or dicyandiamide (DCD): (a) N₂O, (b) CH₄, and (c) soil NO₃ in surface 15 cm. Single arrows indicate an irrigation and double arrows indicate a rainfall >25 mm.

was available in all plots at the time of the first irrigation, but during the first 60 DAF, 528 g N_2O-N were emitted from urea-treated plots, and only 116 g N_2O-N from the unfertilized plots (that also had 60 mg NO_3-N kg^{-1}). Nitrous oxide emissions in the spring after the 177-DAF irrigation were small (< 18 kg N ha^{-1} d^{-1}), and no treatment differences were apparent ($P >$ 0.05). The results of a total denitrification study conduced after the irrigation at 230 DAF (Bronson & Mosier, 1991) indicated that at soil moistures near field capacity, high rates of denitrification occurred, but that the product was almost entirely N_2. Nitrous oxide was only about 2% of the total NO_3 denitrified. The spring N_2O fluxes, therefore, were probably due to denitrification of the NO_3 that was still present in all plots, and represent a much smaller amount of NO_3 reduced than other sites with lighter-textured soils.

Methane Consumption

Consumption rates of CH_4 in irrigated-wheat soil were about one-half those in dryland-wheat soil, but likewise increased as the growing season progressed (Fig. 9–2b). The highest CH_4 consumption rates were observed from the unfertilized control, urea alone, and urea plus DCD. Methane consumption was significantly inhibited by C_2H_2 ($P > 0.05$) in this study, as has been reported in pure cultures of methanotrophic bacteria (Dalton, 1977), but not by DCD ($P > 0.05$). Other studies have indicated that NH_4 fertilizers inhibit CH_4 oxidation in pure culture (Ferenci et al., 1975) and in soil (Steudler et al., 1989; Mosier et al., 1991), but this was not observed in this study.

Irrigated Corn

Nitrous Oxide Emissions

For the first 7 DAF in the ECC-treated plots, blockage of the N_2O to N_2 reduction step of denitrification by C_2H_2 (Yoshinari et al., 1977) was observed (Fig. 9–3a). Laboratory studies (Bronson et al., 1992) indicated that in this soil maximum blockage of N_2O reduction to N_2 occurs with C_2H_2 concentrations of 1000 μL L^{-1}. Nitrous oxide flux with ECC alone (Bronson et al., 1992) was not different from the high level of ECC with urea during the first 6 DAF ($P > 0.05$). This indicates that N_2O emissions from the ECC-treated plots during the 7 DAF (when C_2H_2 production was high) came from denitrification of residual soil NO_3 (6 mg N kg^{-1}) and not from fertilizer N. Urea alone resulted in the highest N_2O emissions for most of the growing season ($P < 0.05$). Nitrous oxide fluxes peaked at 32 DAF, two d after the second irrigation. Flux from urea-treated soil without NIs was 246 g N ha^{-1} d^{-1}, urea plus nitrapyrin emitted 116 g N ha^{-1} d^{-1}, and N_2O fluxes from the other plots were less than 3 g N ha^{-1} d^{-1} (three groups are significantly different at $P < 0.01$). Nitrous oxide emissions declined after the hailstorm at 32 DAF, and since no further irrigations were applied, N_2O fluxes became negligible by 62 DAF.

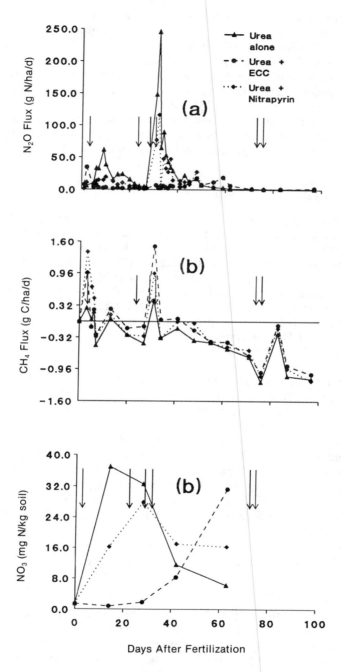

Fig. 9–3. Trace gas fluxes from irrigated corn as affected by urea with and without encapsulated calcium carbide (ECC) or nitrapyrin: (a) N_2O, (b) CH_4, and (c) soil NO_3 in surface 25 cm. Single arrows indicate an irrigation and double arrows indicate a rainfall >25 mm.

Nitrous oxide fluxes were positively correlated with NO_3–N in the surface 30 cm soil ($r = 0.38$, $P < 0.01$). By delaying and reducing NO_3 formation (Fig. 9–3c), the NIs indirectly reduced N_2O emissions from denitrification. Part of the N_2O emitted from urea in the absence of nitrification inhibitors was probably evolved during nitrification of NH_4 (Bremner & Blackmer, 1978). The NH_4–N levels in soil were not correlated with N_2O fluxes, ($r = 0.06$, $P > 0.05$), however.

Total denitrification measurements ($N_2 + N_2O$) (Bronson et al., 1992) indicated that when denitrification activity was high, the ratio of N_2O–($N_2 + N_2O$) was less than 0.3, and when activity was low, N_2O–($N_2 + N_2O$) was 0.5.

Methane Consumption

Methane emissions were observed for brief periods following irrigations on 3 and 29 DAF (Fig. 9–3b), but no significant treatment differences ($P > 0.05$) existed. Low rates of CH_4 consumption predominated during the rest of the growing season. Overall, rates of CH_4 consumption were about two-thirds those in the irrigated wheat field. The highest CH_4 consumption rates were observed from the unfertilized control and the urea-treated plots between the two irrigations. Only at 7 DAF was CH_4 consumption with urea alone significantly less than with the control plots ($P < 0.05$). After the second irrigation, CH_4 consumption with urea alone was similar to that of the unfertilized soils ($P > 0.05$). Nitrification and plant consumption depleted NH_4 by this time.

Acetylene significantly inhibited CH_4 consumption ($P < 0.05$) similar to the irrigated wheat (this paper). In agreement with Dalton's (1977) work in pure cultures of methanotrophic bacteria, nitrapyrin inhibited CH_4 consumption. Acetylene produced in the soil from ECC with or without N fertilizer (data for ECC alone not shown) inhibited CH_4 consumption (compared to control and urea alone) up to 62 DAF, after which no treatment differences in CH_4 consumption were observed. Nitrapyrin had an inhibitory effect on CH_4 consumption ($P < 0.05$), but not on as many sampling dates as C_2H_2 had. Methane consumption generally increased as the growing season progressed, probably due to the disappearance of N fertilizer and NIs.

Cumulative Fluxes of Nitrous Oxide and Methane

Dryland versus Irrigated Wheat

Cumulative emissions of N_2O from unfertilized dryland winter wheat were about one-fourth of that of the unfertilized irrigated wheat (Table 9–1). This is probably due to the lower amounts of inorganic N in the dryland soil.

During the growing season there is no statistical difference between the cumulative CH_4 consumed from the wheat-cropped vs. the fallow field ($P > 0.05$). The wheat-cropped soil did, however, consume more towards the end of the season (Fig. 9–1b; $P < 0.05$) when crop water use dried the soil

Table 9-1. Cumulative fluxes of N$_2$O–N and CH$_4$–C in dryland wheat–fallow system[†], irrigated wheat[‡], and irrigated corn[§].

System-treatment	N$_2$O Avg.	SD	CH$_4$ Avg.	SD
	g N$_2$O-N ha^{-1}		g CH$_4$-C ha^{-1}	
Dryland wheat				
Wheat	101 a[¶]	15	−393 a[¶]	135
Fallow	112 a	18	−257 a	125
Irrigated wheat				
Urea alone	929 a[#]	150	−188 b[#]	22
Urea + ECC (20 kg ha^{-1})	509 b	67	−66 a	11
Urea + DCD (10% N)	437 b	71	−156 b	20
ECC alone (20 kg ha^{-1})	360 b	88	−75 a	6
Control	440 b	102	−185 b	21
Irrigated corn				
Urea alone	1651 a[#]	618	−43 b[#]	7
Urea + nitrapyrin (0.5 L ha^{-1})	980 b	593	−36 ab	9
Urea + ECC (20 kg ha^{-1})	483cc	128	−25 a	8
Control	108 c	29	−45 b	2

[†] Fluxes were measured from time of planting to just before harvest (329 d after planting).

[‡] Fluxes were measured from time of fertilization (planting) to just before harvest (292 d after fertilization).

[§] Fluxes were measured from time of fertilization (9 wk after planting) to just before harvest (97 d after fertilization).

[¶] Means in a column followed by the same letter are not significantly different at $P = 0.05$ by T test.

[#] Means in a column followed by the same letter are not significantly different at $P = 0.05$ by Duncans Mean Range Test.

relative to the fallow field (Fig. 9–1c). Dryland wheat soil consumed twice as much CH$_4$ as unfertilized irrigated wheat ($P < 0.05$). This is probably a result of wetter conditions and the fertilization history of the irrigated wheat field, which would tend to suppress CH$_4$ consumption.

Urea fertilizer addition to irrigated wheat resulted in two times more N$_2$O-N emissions than the unfertilized control (Table 9–1). Severalfold increases of N$_2$O flux with N fertilizer application have been reported often (Mosier et al., 1986; Rolston et al., 1976). Cumulative N$_2$O emitted during the growing season was 929 g N$_2$o-N ha^{-1} from urea alone, or 0.3% applied fertilizer N (after subtracting the cumulative flux of the control). The NIs ECC and DCD reduced N$_2$O emissions 45 and 53%, respectively.

The clay soil under irrigated wheat acted as a small sink for atmospheric CH$_4$. Methane consumption averaged 186 g CH$_4$-C ha^{-1} in both unfertilized and urea-fertilized soil during the growing season. This result, that N fertilizer additions to a cropped system did not affect CH$_4$ consumption, differs from reports in forest soils (Steudler et al., 1989) and in grasslands (Mosier et al., 1991). Several years of N fertilization might have suppressed CH$_4$ consumption to the low levels observed in this study, such that the N additions in this experiment had no effect.

On average, the CH_4 consumption rates observed in the irrigated wheat field were about four times less than that measured in nearby grasslands that were plowed 50 yr ago, and about eight times less than nearby native grasslands that have never been plowed (Ojima et al., 1992). Cultivation itself might have a suppressive effect on CH_4 consumption in soils, as Keller et al. (1990) reported that cropped–cultivated tropical soils had CH_4 consumption rates four times less than that of tropical forest soils.

In the irrigated wheat, C_2H_2 produced from ECC inhibited cumulative CH_4 consumption 65% during the 300-d period. Dicyandiamide addition had no effect on CH_4 consumption ($P > 0.05$), but since it inhibited nitrification (Fig. 9–2c), it was found to be a differential inhibitor. At 168 DAF, DCD had inhibited nitrification 83%, but the cumulative CH_4 flux up to that time was inhibited only 8% (not significant at $P = 0.05$).

Irrigated Corn

Cumulative CH_4 and N_2O fluxes are presented in Table 9–1. This clay loam soil under irrigated corn acted as a small net sink for atmospheric CH_4. As in the irrigated wheat, addition of urea did not affect CH_4 consumption ($P > 0.05$). About 44 g CH_4–C ha^{-1} were taken up by unfertilized and urea-fertilized soil during the growing season. Although the growing season for irrigated corn was one-third of that for irrigated wheat, the cumulative CH_4 consumed in the irrigated corn was about one-fourth of that in the irrigated wheat.

Acetylene produced from ECC inhibited cumulative CH_4 consumption 43% during the 97-d period. Nitrapyrin addition to urea reduced cumulative CH_4 consumption 22% ($P < 0.10$), in agreement with the work of Topp and Knowles (1984) with pure cultures of methanotrophs.

Cumulative N_2O emitted during the growing season was 1480 g N_2O–N ha^{-1} from urea alone, or 0.7% applied fertilizer N. Dinitrogen is the main denitrification product in this clay loam soil (36% clay), but it produces more N_2O per NO_3 denitrified than does the clay soil planted to irrigated wheat. Encapsulated calcium carbide controlled N_2O emissions as effectively as nitrapyrin ($P > 0.05$). Urea fertilizer resulted in 13 times more N_2O–N emissions than the unfertilized control.

SUMMARY

Nitrous oxide emissions increased with increased soil moisture and N fertilization; however, potential for N_2O production from denitrification was limited in heavy clay soils because of enhanced N_2 production. Nitrification inhibitors reduced N_2O emissions in N fertilized, irrigated wheat and corn. Methane consumption was higher in unfertilized dryland wheat than in fertilized, irrigated systems. Urea fertilization in irrigated crops did not affect CH_4 consumption, but the nitrification inhibitors ECC and nitrapyrin generally had a repressive effect. Considering the 10- and 150-yr decay times of CH$_4$ and N$_2$O, respectively, in the atmosphere (Rodhe, 1990), the emissions of N$_2$O contribute far more to the "greenhouse effect" than does the consumption of CH$_4$ in all of these cropping systems. Therefore, strategies to reduce N$_2$O emissions should be emphasized, especially in systems where CH$_4$ consumption rates are low.

REFERENCES

Aulakh, M.S., D.A. Rennie, and E.A. Paul. 1984. Acetylene and N-Serve effects upon N$_2$O emissions from NH$_4^+$ and NO$_3^-$ treated soils under aerobic and anaerobic conditions. Soil Biol. Biochem. 16:351–356.

Aulakh, M.S., J.W. Doran, and A.R. Mosier. 1991. Field evaluation of four methods of measuring denitrification. Soil Sci. Soc. Am. J. 55:1332–1338.

Banerjee, N.K., and A.R. Mosier. 1989. Coated calcium carbide as nitrification inhibitor in upland and flooded soils. J. Indian Soc. Soil Sci. 37:306–313.

Bremner, J.M., and A.M. Blackmer. 1978. Nitrous oxide: Emission from soils during nitrification of fertilizer nitrogen. Science (Washington, DC) 199:295–296.

Bronson, K.F., and A.R. Mosier. 1991. Effect of encapsulated calcium carbide and dicyandiamide on N$_2$O and CH$_4$ fluxes in irrigated wheat. p. 259. In Agronomy abstracts. ASA, Madison, WI.

Bronson, K.F., A.R. Mosier, and S.R. Bishnoi. 1992. Nitrous oxide emissions in irrigated corn as affected by nitrification inhibitors. Soil Sci. Soc. Am. J. 56:161–165.

Bouwman, A.F. 1990. Soils and the greenhouse effect. John Wiley & Sons, Chichester, England.

Crutzen, P.J. 1976. Upper limits on atmospheric ozone reductions following increased application of fixed nitrogen to the soil. Geophys. Res. Lett. 3:169–172.

Dalton, H. 1977. Ammonia oxidation by the methane oxidising bacterium Methylococcus capsulatus Strain Bath. Arch. Microbiol. 114:273–279.

Eichner, M.J. 1990. Nitrous oxide emissions from fertilized soils: Summary of available data. J. Environ. Qual. 19:272–280.

Ferenci, T., T. Strom, and J.R. Quale. 1975. Oxidation of carbon monoxide and methane by Pseudomonas methanica. J. Gen. Microbiol. 91:79–91.

Hutchinson, G.L., and A.R. Mosier. 1981. Improved soil cover method for field measurement of nitrous oxide fluxes. Soil Sci. Soc. Am. J. 45:311–316.

Jones, R.D., and R.Y. Morita. 1983. Methane oxidation by Nitrosococcus oceanus and Nitrosomonas europaea. Appl. Environ. Microbiol. 45:401–410.

Keller, M., M.E. Mitre, and R.F. Stallard. 1990. Consumption of atmospheric methane in soils of central Panama: Effects of agricultural development. Global Biogeochem. Cycles 4:21–27.

Megalhaes, A., P.M. Chalk, and W.M. Strong. 1984. Effect of nitrapyrin and nitrous oxide emission from fallow soils fertilized with anhydrous ammonia. Fert. Res. 5:411–421.

Mosier, A.R., and L. Mack. 1980. Gas chromatographic system for precise, rapid analysis of N$_2$O. Soil Sci. Soc. Am. J. 44:1121–1123.

Mosier, A.R., W.D. Guenzi, and E.E. Schweizer. 1986. Soil losses of dinitrogen and nitrous oxide from irrigated crops in Northeastern Colorado. Soil Sci. Soc. Am. J. 50:344–348.

Mosier, A., D. Schimel, D. Valentine, K. Bronson, and W. Parton. 1991. Methane and nitrous oxide fluxes in native, fertilized and cultivated grasslands. Nature (London) 350:330–332.

Ojima, D.S., D.W. Valentine, A.R. Mosier, W.J. Parton, and D.S. Schimel. 1992. Effect of land use change on soil methane oxidation in temperate forest and grassland soils. Chemosphere. (in press.)

Rodhe, H. 1990. A comparison of the contribution of various gases to the greenhouse effect. Science (Washington, DC) 248:1217–1219.

Rolston, D.E., M. Fried, and D.A. Goldhamer-Seiler, W., and R. Conrad. 1987. Contribution of tropical ecosystems to the global budgets of trace gases, especially CH$_4$, H$_2$, CO and N$_2$O. p. 133–160. In R.E. Dickinson (ed.) Geophysiology of Amazonia. Vegetation and climate interactions, John Wiley & Sons, New York.

Steudler, P.A., R.D. Bowden, J.M. Melillo, and J.D. Aber. 1989. Influence of nitrogen fertilization on methane uptake in temperate forest soils. Nature (London) 341:314–316.

Topp, E., and R. Knowles. 1984. Effects of nitrapyrin [2-chloro-6-(trichloromethyl)pyridine] on the obligate methanotroph methylosinus trichosporium OB3b. Appl. Environ. Microbiol. 47:258–262.

Whalen, S.C., and W.S. Reeburgh. 1990. Consumption of atmospheric methane by tundra soils. Nature (London) 346:160–162.

Whittenbury, R., K.C. Phillips, and J.F. Wilkinson. 1970. Enrichment, isolation and some properties of methane-utilizing bacteria. J. Gen. Microbiol. 61:205–218.

Yoshinari, T., R. Hynes, and R. Knowles. 1977. Acetylene inhibition of nitrous oxide reduction and measurement of denitrification and nitrogen fixation in soil. Soil Biol. Biochem. 9:177–183.

10   Methane: Processes of Production and Consumption

Roger Knowles

Department of Microbiology
McGill University
Sainte Anne de Bellevue, Canada

Emissions of methane (CH_4) from environments in the field, whether terrestrial or aquatic, natural or agricultural, depend on a multitude of factors not least of which are the biological processes by which CH_4 is produced and consumed. The net emission from, say, an agricultural system is the result of production (methanogenesis) and consumption (methanotrophy), and the net emission will be positive or negative depending on the relative magnitudes of these processes. It is therefore important to distinguish between them, to estimate their individual rates, and to understand how they are regulated by nutritional, environmental and perhaps other factors. This article summarizes some of these aspects.

METHANE PRODUCTION

Nutrition

Organisms with the ability to produce CH_4 are restricted to the archaeobacteria, a group of bacteria that often live in very extreme environments and which are distinct, in many ways, from the regular eubacteria and the eukaryotes. Methanogens are important because they represent the terminal steps in most anaerobic food webs (the rumen, aquatic sediments, and waterlogged soils) where organic C is ultimately released as CH_4. In such systems, the hydrolysis and fermentation of biological polymers and other molecules results in the production of many intermediates and products, including fatty acids, alcohols, and gaseous hydrogen (H_2) and carbon dioxide (CO_2) (Fig. 10–1) (Wolin & Miller, 1987).

Methanogenic bacteria can use many of these products, including H_2, as their substrate. Apparently 77% of 68 described species are hydrogenotrophs and most of these can also use formate; 14% use acetic acid, and 28% are methylotrophs, using 1-C compounds such as methanol or methylated

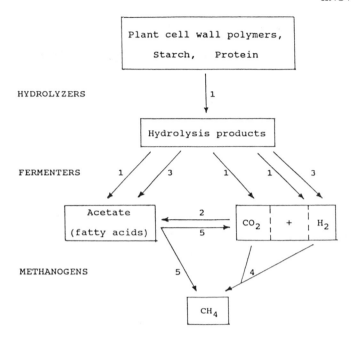

1. Fermenters   4. Methanogens using $H_2 + CO_2$
2. H_2-consuming acetogens   5. Methanogens using acetate
3. Acetogens producing H_2 and producing CO_2

Fig. 10-1. Some of the pathways by which plant residues are converted to CH_4.

amines (Garcia, 1990). The overall reactions are listed in Table 10-1. Most can fix CO_2 and most need H_2 as their electron donor (energy source) even if they use acetate as a C source.

In nature, the major substrates for CH_4 production are acetate and H_2-CO_2. It is possible that autotrophy by these bacteria is important in nature (Zeikus, 1977). Even if other conditions are favorable, methanogens are probably substrate-limited in nature, and probably the best-documented example if this is the competition for H_2 (and perhaps acetate and formate, all referred to as competitive substrates) that occurs between methanogens

Table 10-1. Substrates used by methanogens.†

H₂ and CO₂	$4H_2 + CO_2 \text{----------} > CH_4 + 2H_2O$
Formate	$4HCOOH \text{----------} > CH_4 + 3CO_2 + 2H_2O$
Methanol	$4CH_3OH \text{----------} > 3CH_4 + CO_2 + 2H_2O$
Methanol and H₂	$CH_3OH + H_2 \text{-------} > CH_4 + H_2O$
Methylamine	$4CH_3NH_2Cl + 2H_2O \text{----} > 3CH_4 + CO_2 + 4NH_4Cl$
Dimethylamine	$2(CH_3)_2NHCl + 2H_2O \text{---} > 3CH_4 + CO_2 + 2NH_4Cl$
Trimethylamine	$4(CH_3)_3NCl + 6H_2O \text{----} > 9CH_4 + 3CO_2 + 4NH_4Cl$
Acetate	$CH_3COOH \text{----------} > CH_4 + CO_2$

† From Wolin and Miller (1987).

and sulfate-reducing bacteria (Cicerone & Oremland, 1988). This is reported mainly from marine systems but such competition could also exist in anaerobic soils rich in sulfate (SO_4^{2-}). The consumption of H_2 by the methanogens is often important in maintaining low enough H_2 partial pressures to permit active growth of acetogenic bacteria that produce H_2, yet are inhibited by its accumulation. This phenomenon of "interspecies H_2 transfer" is important in many anaerobic ecosystems (Wolin & Miller, 1987).

We have observed in an anaerobic peat that no H_2 accumulates during active methanogenesis, but if the methanogens are inhibited by bromoethanesulfonate (BES), a specific inhibitor of methanogenesis (see later), then H_2 accumulates. This suggests that H_2 is an important and perhaps limiting energy source for the CH_4-producing bacteria in this system. This conclusion is further supported by the fact that for several peat types studied, the addition of H_2 to anaerobic peat stimulates CH_4 production (Dunfield & Knowles, 1991, unpublished data).

Most methanogens use H_2S as their S source (Zeikus, 1977) but whether this is their major source in nature is not clear. Some require and/or are stimulated by vitamins that could also be significant growth factors in nature. All species reportedly use NH_4^+-N as their dominant N source, more oxidized forms being generally absent in anoxic environments at low redox potentials. The reports of N_2 fixation by methanogens are interesting (Belay et al., 1984; Murray & Zinder, 1984) but, so far as we are aware, there has been no conclusive demonstration of their nitrogenase activity in natural anoxic soils or sediments.

Environmental Factors

Methanogenic bacteria are very strict anaerobes and intolerant of O_2 exposure for growth, although they may not be killed by some exposure. They require very low redox potentials (Eh $= < -300$ mV) associated with rapid decomposition of organic matter and the presence of reducing agents such as H_2S or cysteine. When anaerobic peats or paddy soil samples are placed under anaerobiosis in the laboratory, they frequently produce CH_4 without a lag (Knowles, 1991, unpublished data; Mayer & Conrad, 1990). In other more aerobic soils, CH_4 production occurs after a sometimes lengthy lag period (Megraw & Knowles, 1987a; Mayer & Conrad, 1990), which suggests, nevertheless, that methanogenic organisms are indeed present in such habitats.

Temperature is an important regulator of methanogenesis, as of other biochemical reactions. In nature, CH_4 producers appear generally to be mesophiles although thermophilic organisms may appear in certain systems and there is considerable interest in the practical applications of very thermophilic systems for waste disposal and biogas production. We have found, for a limited number of northern peats, that the temperature optimum for methanogenesis is about 20 to 25 °C. The temperature quotients (Q_{10} values) were >5.3, activation energies were in the range 120 to 270 kJ mol^{-1} (see also similar reports of Schütz et al., 1990; Sextone & Mains, 1990), and there was very little if any activity at temperatures of 0 to 10 °C (Dunfield et al., 1992). Also working with northern peats, Svensson (1984) observed that acetate enrichments showed an optimum of 20 °C but H_2-oxidizing enrichments were most active at 28 °C. There appear to be no reports of soils or isolated organisms, even from arctic environments, showing psychrophilic methanogenesis.

Most methanogenic bacteria grow in the pH range 6 to 8 and the lowest pH for growth of any of the species listed by Garcia (1990) is 5.6. Most anaerobic soils will fall within such a range of tolerance. However, the fact that more acidic peats show methanogenic activity (e.g., Moore & Knowles, 1987, 1989, 1990; Moore et al., 1990) suggests that acid tolerant, if not acidophilic, methanogens may occur. The only reported acid-tolerant, isolated methanogen is that of Williams and Crawford (1985) that grew in pure culture at $\geq pH$ 5.3 and produced CH_4 at $\geq pH$ 3.1.

METHANE CONSUMPTION

Nutrition

Bacteria that are able to grow using CH_4 are referred to as methanotrophs and are part of a larger grouping of organisms (termed methylotrophs) that can utilize one-carbon (1-C) compounds having no C–C bonds (Anthony, 1986). There is evidence that certain, mostly SO_4^{2-}-reducing, habitats exist in which anaerobic CH_4 oxidation occurs (Hanson, 1980; Cicerone & Oremland, 1988). However, the organisms responsible have not to our knowledge been isolated and it seems unlikely that this process occurs in soils, although it may occur in anaerobic peat. All methanotrophs isolated and studied to date are obligate aerobes since the enzyme responsible for the initial step in CH_4 oxidation is a monooxygenase enzyme that requires molecular O_2. The product of this reaction, methanol, is further successively oxidized via formaldehyde to formate and then CO_2 (Fig. 10–2). There is some evidence that some of these intermediates may leak or be excreted from cells and perhaps support growth of other bacteria (Ivanova & Nesterov, 1988). The methane monooxygenase (MMO) requires reducing power in the form of NADH or electrons derived from the later steps of CH_4 oxidation (Fig. 10–2). In some methanotrophs, at least, the nature and location of the MMO enzyme depends on the availability of Cu to the cells. In limiting Cu a soluble MMO (sMMO) is formed, but under Cu sufficiency the MMO is particulate (pMMO) (Bédard & Knowles, 1989), and this difference affects the substrate specificity of the enzyme. One sMMO studied has a much greater substrate versatility than the pMMO (Anthony, 1986) and it can cooxidize a wide variety of hydrocarbons, halogenated aliphatics, and related compounds (Table 10–2); thus giving these reactions industrial and bioremediation potential (Henson et al., 1988; Lanzarone & McCarty, 1990). However, since the product is not methanol, they do not provide energy or C for growth. The sMMO has been purified and is well studied but the pMMO is more

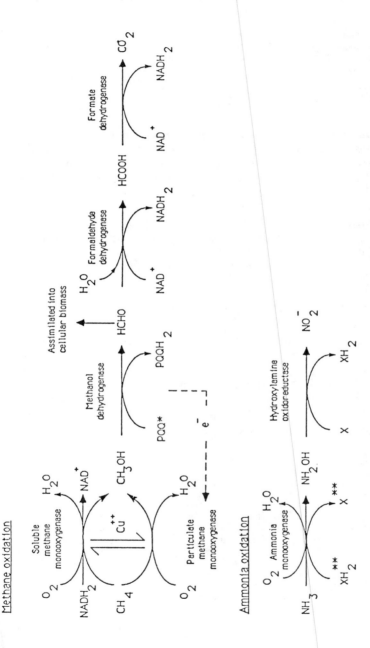

Fig. 10-2. The pathways of CH_4 oxidation in methanotrophs and NH_3 oxidation in NH_3 oxidizers. PQQ = Pyrroquinoline quinone. X and XH_2 are the oxidized and reduced forms of an unknown electron donor (reproduced from Bédard & Knowles [1989] with permission).

Table 10-2. Substrates of sMMO from *Methylococcus capsulatus* (Bath).†

Chloromethane	Ethane	Ethene	Cyclohexane
Bromomethane	Propane	Propene	Benzyne
Iodomethane	Butane	But-1-ene	Toluene
Dichloromethane	Pentane	Dimethylether	Styrene
Trichloromethane	Hexane	Diethylether	Pyridine
Cyanomethane	Heptane		
Nitromethane	Octane	Methanethiol	CO

† Modified from Anthony (1986).

recalcitrant to biochemical analysis. It is not known in which form the MMO exists in any natural environment and the significance of Cu availability and the two forms of the enzyme for the ecology of the bacteria is not clear (Bédard & Knowles, 1989). The obligate methanotrophs are placed in groups (Topp & Hanson, 1991) according to the appearance of their membrane assemblies, the nature of resting stages and other properties (Table 10-3). It is noteworthy that at least some fix CO_2 autotrophically and some fix N_2 (Murrell, 1988; Murrell & Dalton, 1983), but there is no information about such activities in the field. *Methylococcus capsulatus* is more thermotolerant than other organisms. Most grow on ammonia but prefer NO_3^- as the N source.

Affinity of Methanotrophs for Methane

The affinity of a methanotroph for CH_4 governs its ability to compete for CH_4 at low concentrations. Reported half-saturation constants (apparent K_m or K_s values) for CH_4 range from 1 to 160 μM for pure cultures, purified MMOs, and environmental samples (Megraw & Knowles, 1987a; Whalen et al., 1990). A recent estimate for peat from our experiments is 2 μM (Dunfield et al., 1992). Thus, all of the reported values are too high to support the growth of methanotrophs on the small concentrations (approximately 2.5 nM at 20 °C, that is, about three orders of magnitude lower) dis-

Table 10-3. Characteristics of major groups of methanotrophs.†

	Type I	Type X	Type II
Membranes	Bundles of vesicular disks		Paired at periphery of cell
Resting stages	Cysts (*Azotobacter*-like)		Exospores ("lipid cyst")
C assimilation pathway	Ribulose monophosphate pathway		Serine pathway
TCA cycle	Lacks 2-oxoglutarate dehydrogenase		Complete
Fix CO_2 via rubisco	−	+	−
Fix N_2	?	+	+
Cell shape	Rod	Coccus	Rod and vibrio
Growth at 45 °C	Some +	+	−
Examples of genera	*Methylomonas*	*Methylococcus*	*Methylosinus* *Methylobacterium* (uses glucose)

† Modified and abbreviated from Whittenbury and Krieg (1984).

solved in an aqueous phase in equilibrium with the atmosphere. Some soils act as net CH_4 sinks (one of the first was reported by Harris et al. [1982] in the Great Dismal Swamp area) but no organisms showing very low apparent K_m values appear to have been isolated from these soils.

Environmental Factors

Methanotrophic bacteria are necessarily aerobic (O_2 being essential for the MMO), although some prefer O_2 concentrations below ambient. Therefore they occur and are active close to oxic–anoxic interfaces in nature where the concentration gradients of CH_4 and O_2 overlap. Indeed active methanotrophs that have a relatively high affinity for O_2 can be responsible for its depletion and consequent cessation of activity of autotrophic nitrifiers, with which they are in competition for O_2 (Megraw & Knowles, 1987b). The study of methanotrophs in generally aerobic soils is in its infancy at present, but their existence suggests either that sufficient CH_4 is available periodically to assure their survival, or that the bacteria are facultative methanotrophs that are able to utilize other organic compounds for maintenance and perhaps for growth. If methanotrophs are growing in a N-limited environment in which they must fix N_2, they will of course be highly microaerophilic because of the O_2 sensitivity of their nitrogenase, regardless of their O_2 tolerance under conditions of N excess.

Most methanotrophs appear to be mesophiles, but some such as *M. capsulatus* are more tolerant of high temperatures (up to about 50°C). There seem to be no reports of the isolation of psychrophilic organisms, and our studies with northern peats suggest that there is no marked adaptation to low temperature environments. In contrast to our findings with CH_4 production rates mentioned earlier, the temperature quotients (Q_{10}) for consumption were lower (1.4–2.1), activation energies were lower (20–80 kJ mol^{-1}), and there was appreciable (13–38% of maximum) activity at 0 to 5°C (Dunfield et al., 1992). This behavior will accentuate any decrease in emissions at low temperatures in the field.

Consumption of CH_4 by some peat soils occurs even at pH values below 4.0 (Moore & Knowles, 1990) but no isolations are reported from such environments. Methanotrophic yeasts (*Rhodotorula* and *Candida* spp.) tolerant to pH's of 4.4 and 3.8, were isolated by Wolf and Hanson (1980) and Saha and Sen (1989), respectively, and enrichment cultures of bacteria were obtained by Heyer and Suckow (1985) from peat and peat water samples down to pH 5.0 and 4.4, respectively. They were grown in media of pH 6.4 to 6.5 and their pH response was not tested.

Nitrification by Methanotrophs

The broad substrate range of the sMMO as well as the potential of methanotrophs for cometabolism and degradative remediation has already been mentioned (e.g., Henson et al., 1988; Lanzarone & McCarty, 1990). Another activity of some potential significance in N biogeochemistry is the ability of the MMO to cooxidize ammonia and so contribute to nitrification (see Bédard & Knowles, 1989; Topp & Hanson, 1991, and references therein). The MMO of methanotrophs and the ammonia monoxygenase (AMO) of nitrifiers have similar substrate specificities and it appears that CH_4 and NH_3 are competitive substrates of both enzymes. It seems that the product of NH_3 oxidation by MMO is hydroxylamine (NH_2OH, Fig. 10-2), which is then further oxidized by hydroxylamine oxidoreductase (distinct from methanol dehydrogenase) to NO_2^- (Bédard & Knowles, 1989). Oxidation of NO_2^- to NO_3^- does not seem to occur normally in methanotrophs but this reaction could be catalyzed by associated autotrophic or heterotrophic nitrifiers acting in a consortium (e.g., Megraw & Knowles, 1989). Thus the autotrophic nitrifiers and the methanotrophs occur in similar habitats, and may compete for O_2, CH_4 and NH_3 (Bédard & Knowles, 1989). This may be a reason for the reported inhibitory effects of added NH_3 on CH_4 consumption in some forest (Steudler et al., 1989) and grassland (Mosier et al., 1991) soils. Methanotrophs can also participate in gaseous N oxide metabolism— pure cultures of organisms tested produce but do not consume N_2O (Yoshinari, 1985; Krämer et al., 1990; Knowles & Topp, 1988), and consume but do not produce NO (Krämer et al., 1990). It may be necessary to take this into account in evaluating gaseous N oxide emissions in field studies.

INHIBITORS AS TOOLS

In experimental work on CH_4 metabolism it is often desirable to attempt to inhibit specifically one process to determine the effect of the process on the entire system. An exhaustive review of this topic is not intended here (the reader is referred to Bédard & Knowles [1989], and Oremland & Capone [1988] for more details). Some compounds that have been used or that have potential for inhibiting CH_4 production and/or consumption are shown in Tables 10-4 and 10-5. It may be noted that nitrapyrin and acetylene, present on both lists, also inhibit the AMO of autotrophic nitrifiers (Bédard & Knowles, 1989). Sulfate and molybdate, although not directly affecting methanogens, may influence their activity indirectly. Sulfate promotes SO_4^{2-}-reducing bacteria and enhances their competition for H_2, and molybdate, by inhibiting SO_4^{2-} reducers, may indirectly stimulate methanogens. However, molybdate may inhibit methanogens by precipitating H_2S, which is required by many methanogens (Table 10-4). The BES is highly specific since it is an analogue of the methanogen cofactor CoM. Ethylene is also specific and, at least in anaerobic systems, does not appear to be metabolized (Schink, 1985).

Many of the compounds listed in Table 10-5 are MMO substrates (NH_3, dimethylether, 1,2-epoxypropane), suicide substrate (acetylene), or analogue (methylfluoride), but the relative sensitivities to these compounds of the two forms of enzyme, sMMO and pMMO, has not been carefully examined. Methylfluoride at 10% v v $^{-1}$ in the gas phase (Table 10-5) and even at 1% v v $^{-1}$ (Oremland & Culbertson, 1992) inhibits methanotrophic activity, but

Table 10-4. Some inhibitors of methanogenesis.

Compound	Comment	Concentration[†]	References[‡]
		μM	
Bromoethane-sulfonate (BES)	Analogue of mercaptoethanesulfonic acid cofactor (CoM)	100	1, 2
Ethylene	Fairly specific, mechanism unknown	5	3
Acetylene	May cause loss of transmembrane pH gradient	8	4, 5, 6
Nitrapyrin	Mechanism not known	2	7
N oxides	Not redox effect or substrate competition	50 mg N kg^{-1}	8
		1	9
Sulfate	Unclear, may be competition for H_2	--	10
Molybdate	Unclear, may precipitate needed H_2S	--	11

† Minimum concentration reported to be inhibitory.
‡ 1 = Oremland and Capone (1988), 2 = Sparling and Daniels (1987), 3 = Schink (1985), 4 = Sprott et al. (1982), 5 = Raimbault (1975), 6 = Oremland and Taylor (1975), 7 = Salvas and Taylor (1980), 8 = Bollag and Czlonkowsky (1973), 9 = Balderston and Payne (1976), 10 = Cicerone and Oremland (1988), and 11 = Oremland and Capone (1988).

it is not significantly oxidized by *M. capsulatus* (Texas) (Meyers, 1980). Meyers (1980) did not report its effect on CH_4 oxidation. Some agents (allylthiourea, thiourea and perhaps nitrapyrin) are believed to inhibit by acting as chelators, presumably of the Cu in pMMO (Bédard & Knowles, 1989).

It is clear that few of the inhibitors mentioned are highly specific, and, in the case of methanotrophic activities in nature, the role of the sMMO vs. the pMMO in determining inhibitor sensitivity is not yet clear.

Table 10-5. Some inhibitors of CH_4 oxidation.

Compound	Comment	Concentration[†]	References[‡]
Ammonia(um)	Competition as substrate at MMO	K_i 0.2-10 mM§	1, 2
Methyl fluoride	Substrate analogue	10% v v^{-1}	3
Dimethylether	Substrate analogue?	10%	3, 4
1,2-epoxypropane	Alkene oxidation more sensitive	K_i 5 mM	5
Picolinic acid	Action unknown, may chelate Cu	K_i 8 μM	6, 7
Nitrapyrin	Action unknown, may chelate Cu	9 μM	8
Acetylene	Suicide substrate	K_i 0.02 μM	9, 10
Allythiourea	Metal (Cu?) chelator	10 μM	11

† Minimum concentration reported to be inhibitory.
‡ 1 = Ferenci et al. (1975), 2 = O'Neil and Wilkinson (1977), 3 = Oremland and Culbertson (1992), 4 = Meyers (1982), 5 = Habetz-Crützen and de Bont (1985); 6 = Megraw and Knowles (1990), 7 = Salvas and Taylor (1984), 8 = Topp and Knowles (1982, 1984), 9 = de Bont and Mulder (1976), 10 = Prior and Dalton (1985), and 11 = Hubley et al. (1975).
§ K_i = concentration yielding 50% inhibition of activity.

OTHER FACTORS REGULATING METHANE FLUX

Light may have very little effect on production and consumption of CH_4 in most soils, but in wet areas with algal or cyanobacterial surface growth the production of O_2 during the daytime may stimulate CH_4 oxidation by increasing O_2 availability to the methanotrophic bacteria (King, 1990).

Another factor that may be important is the physical nature of the soil profile in controlling its transport resistance or gas permeability. Born et al. (1990) conclude, from measurements of CH_4 concentration gradients and Rn fluxes, that gas permeability exerts greater control on CH_4 uptake rates (negative fluxes) than does the potential CH_4 oxidation rate. Furthermore, in many wetland and other environments the entire methanotrophic region in the soil may be bypassed by gas transport pathways through higher plants such as rice (*Oryza sativa* L.) and others possessing lacunae (Holzapfel-Pschorn et al., 1985; Nouchi et al., 1990) and perhaps also plants without such gas spaces (Whalen & Reeburgh, 1990). In these cases the CH_4-consuming organisms at the soil surface may have little impact. Rhizosphere methanotrophs would, of course, continue to play a role.

SUMMARY

The processes by which CH_4 is produced by methanogenic bacteria and consumed by methanotrophic bacteria are outlined, and the factors that regulate these processes in nature are discussed.

REFERENCES

Anthony, C. 1986. Bacterial oxidation of methane and methanol. Adv. Microb. Physiol. 27:113–210.

Balderston, W.L., and W.J. Payne. 1976. Inhibition of methanogenesis in salt marsh sediments and whole-cell suspensions of methanogenic bacteria by nitrogen oxides. Appl. Environ. Microbiol. 32:264–269.

Bédard, C., and R. Knowles. 1989. Physiology, biochemistry, and specific inhibitors of CH_4, NH_4^+, and CO oxidation by methanotrophs and nitrifiers. Microbiol. Rev. 53:68–84.

Belay, N., R. Sparling, and L. Daniels. 1984. Dinitrogen fixation by a thermophilic methanogenic bacterium. Nature (London) 312:286–288.

Bollag, J.-M., and S.T. Czlonkowski. 1973. Inhibition of methane formation in soil by various nitrogen-containing compounds. Soil Biol. Biochem. 5:673–678.

Born, M., H. Dörr, and I. Levin. 1990. Methane consumption in aerated soils of the temperate zone. Tellus 42(B):2–8.

Cicerone, R.J., and R.S. Oremland. 1988. Biogeochemical aspects of atmospheric methane. Global Biogeochem. Cycles 2:299–327.

De Bont, J.A.M., and E.G. Mulder. 1976. Invalidity of the acetylene reduction assay in alkane-utilizing, nitrogen-fixing bacteria. Appl. Environ. Microbiol. 31:640–647.

Dunfield, P., R. Knowles, R. Dumont, and T.R. Moore. 1992. Methane production and consumption in temperate and subarctic peat soils: Response to temperature and pH. Soil Biol. Biochem. (In press.)

Ferenci, T., T. Strom, and J.R. Quayle. 1975. Oxidation of carbon monoxide and methane by *Pseudomonas methanica*. J. Gen. Microbiol. 91:79–91.

Garcia, J.L. 1990. Taxonomy and ecology of methanogens. FEMS Microbiol. Rev. 87:287–308.

Habets-Crützen, A.Q.H., and J.A.M. de Bont. 1985. Inactivation of alkene oxidation by epoxides in alkene- and alkane-grown bacteria. Appl. Microbiol. Biotechnol. 22:428–433.

Hanson, R.S. 1980. Ecology and diversity of methylotrophic organisms. Adv. Appl. Microbiol. 26:3–39.

Harriss, R.C., D.I. Sebacher, and F.P. Day. 1982. Methane flux in the Great Dismal Swamp. Nature (London) 297:673–674.

Henson, J.M., M.V. Yates, J.W. Cochran, and D.L. Shackleford. 1988. Microbial removal of halogenated methanes, ethanes, and ethylenes in an aerobic soil exposed to methane. FEMS Microbiol. Ecol. 53:193–201.

Heyer, J., and R. Suckow. 1985. Ökologische Untersuchungen der Methanoxidation in einem sauren Moorsee. Limnologica 16:247–266.

Holzapfel-Pschorn, A., R. Conrad, and W. Seiler. 1985. Production, oxidation and emission of methane in rice paddies. FEMS Microbiol. Ecol. 31:343–351.

Hubley, J.H., A.W. Thomson, and J.F. Wilkinson. 1975. Specific inhibitors of methane oxidation in *Methylosinus trichosporium*. Arch. Microbiol. 102:199–202.

Ivanova, T.I., and A.I. Nesterov. 1988. Production of organic exometabolites by diverse cultures of obligate methanotrophs. Microbiology 57:486–491. Transl. of Mikrobiologiya 57:600–605.

King, G.M. 1990. Regulation by light of methane emissions from a wetland. Nature (London) 345:513–515.

Knowles, R., and E. Topp. 1988. Some factors affecting nitrification and the production of nitrous oxide by the methanotrophic bacterium *Methylosinus trichosporium* OB3b. p. 383–393. *In* G. Giovannozzi-Sermanni and P. Nannipieri (ed.) Current perspectives in environmental biogeochemistry. Consiglio Nazionale delle Ricerche-I.P.R.A., Rome.

Krämer, M., M. Baumgärtner, M. Bender, and R. Conrad. 1990. Consumption of NO by methanotrophic bacteria in pure culture and in soil. FEMS Microbiol. Ecol. 73:345–350.

Lanzarone, N.A., and P.L. McCarty. 1990. Column studies on methanotrophic degradation of trichloroethene and 1,2-dichloroethane. Ground Water 28:910–919.

Mayer, H.P., and R. Conrad. 1990. Factors influencing the population of methanogenic bacteria and the initiation of methane production upon flooding of paddy soil. FEMS Microbiol. Ecol. 73:103–112.

Megraw, S., and R. Knowles. 1987a. Methane production and consumption in a cultivated humisol. Biol. Fertil. Soils 5:56–60.

Megraw, S., and R. Knowles. 1987b. Active methanotrophs suppress nitrification in a humisol. Biol. Fertil. Soil 4:205–212.

Megraw, S., and R. Knowles. 1989. Methane-dependent nitrate production by a microbial consortium enriched from a cultivated humisol. FEMS Microbiol. Ecol. 62:359–366.

Megraw, S.R., and R. Knowles. 1990. Effect of picolinic acid (2-pyridine carboxylic acid) on the oxidation of methane and ammonia in soil and in liquid culture. Soil Biol. Biochem. 22:635–641.

Meyers, A.J. 1980. Evaluation of bromomethane as a suitable analogue in methane oxidation studies. FEMS Microbiol. Lett. 9:297–300.

Meyers, A.J. 1982. Obligate methylotrophy: Evaluation of dimethylether as a C-1 compound. J. Bacteriol. 150:966–968.

Moore, T.R., and R. Knowles. 1987. Methane and carbon dioxide evolution from subarctic fens. Can. J. Soil Sci. 67:77–81.

Moore, T.R., and R. Knowles. 1989. The influence of water table levels on methane and carbon dioxide emissions from peatland soils. Can. J. Soil Sci. 69:33–38.

Moore,T.R., and R. Knowles. 1990. Methane emissions from fen, bog, and swamp peatlands in Québec. Biogeochemistry 11:45–61.

Moore, T., N. Roulet, and R. Knowles. 1990. Spatial and temporal variations of methane flux from subarctic/northern boreal fens. Global Biogeochem. Cycles 4:29–46.

Mosier, A., D. Schimel, D. Valentine, K. Bronson, and W. Parton. 1991. Methane and nitrous oxide fluxes in native, fertilized and cultivated grasslands. Nature (London) 350:330–332.

Murray, P.A., and S.H. Zinder. 1984. Nitrogen fixation by a methanogenic archaebacterium. Nature (London) 312:284–286.

Murrell, J.C. 1988. The rapid switch-off of nitrogenase activity in obligate methane-oxidizing bacteria. Arch. Microbiol. 150:489–495.

Murrell, J.C., and H. Dalton. 1983. Nitrogen fixation in obligate methanotrophs. J. Gen. Microbiol. 129:3481–3486.

Nouchi, I., S. Mariko, and K. Aoki. 1990. Mechanism of methane transport from the rhizosphere to the atmosphere through rice plants. Plant Physiol. 94:59-66.

O'Neill, J.G., and J.F. Wilkinson. 1977. Oxidation of ammonia by methane-oxidizing bacteria and the effects of ammonia on methane oxidation. J. Gen. Microbiol. 100:407-412.

Oremland, R.S., and D.G. Capone. 1988. Use of "specific" inhibitors in biogeochemistry and microbial ecology. Adv. Microbial. Ecol. 10:285-383.

Oremland, R.S., and C.W. Culbertson. 1992. Evaluation of methyl fluoride and dimethyl ether as inhibitors of aerobic methane oxidation. Appl. Environ. Microbiol. 58:2983-2992.

Prior, S.D., and H. Dalton. 1985. Acetylene as a suicide substrate and active site probe for methane monooxygenase from *Methylococcus capsulatus* (Bath). FEMS Microbiol. Lett. 29:105-109.

Raimbault, M. 1975. Étude de l'influence inhibitrice de l'acétylène sur la formation biologique du méthane dans un sol de rizière. Ann. Microbiol. (Inst. Pasteur) 126A:247-258.

Salvas, P.L., and B.F. Taylor. 1980. Blockage of methanogenesis in marine sediments by the nitrification inhibitor 2-chloro-6-(trichloro-methyl)pyridine (nitrapyrin or N-Serve). Curr. Microbiol. 4:305-308.

Salvas, P.L., and B.F. Taylor. 1984. Effect of pyridine compounds on ammonia oxidation by autotrophic nitrifying bacteria and *Methylosinus trichosporium* OB3b. Curr. Microbiol. 10:53-56.

Saha, V., and M. Sen. 1989. Methane oxidation by *Candida tropicalis*. Natl. Acad. Sci. Lett. (India) 12:373-376.

Schink, B. 1985. Inhibition of methanogenesis by ethylene and other unsaturated hydrocarbons. FEMS Microbiol. Ecol. 31:63-68.

Schütz, H., W. Seiler, and R. Conrad. 1990. Influence of soil temperature on methane emission from rice paddy fields. Biogeochemistry 11:77-95.

Sextone, A.J., and C.N. Mains. 1990. Production of methane and ethylene in organic horizons of spruce forest soils. Soil Biol. Biochem. 22:135-139.

Sparling, R., and L. Daniels. 1987. The specificity of growth inhibition of methanogenic bacteria by bromoethane sulfonate. Can. J. Microbiol. 33:1132-1136.

Sprott, G.D., K.F. Jarrell, K.M. Shaw, and R. Knowles. 1982. Acetylene as an inhibitor of methanogenic bacteria. J. Gen. Microbiol. 128:2453-2462.

Steudler, P.A., R.D. Bowden, J.M. Melillo, and J.D. Aber. 1989. Influence of nitrogen fertilization on methane uptake in temperate forest soils. Nature (London) 341:314-316.

Svensson, B.H. 1984. Different temperature optima for methane formation when enrichments from acid peat are supplemented with acetate or hydrogen. Appl. Environ. Microbiol. 48:389-394.

Topp, E., and R.S. Hanson. 1991. Metabolism of radiatively important trace gases by methane-oxidizing bacteria. p. 71-90. *In* J.E. Rogers and W.B. Whitman (ed.) Microbial production and consumption of greenhouse gases: Methane, nitrogen oxides, and halomethanes. Am. Soc. Microbiol., Washington, DC.

Topp, E., and R. Knowles. 1982. Nitrapyrin inhibits the obligate methylotrophs *Methylosinus trichosporium* and *Methylococcus capsulatus*. FEMS Microbiol. Lett. 14:47-49.

Topp, E., and R. Knowles. 1984. Effects of nitrapyrin [2-chloro-6-(trichloromethyl)pyridine] in the obligate methanotroph *Methylosinus trichosporium* OB3b. Appl. Environ. Microbiol. 47:258-262.

Whalen, S.C., and W.S. Reeburgh. 1990. Consumption of atmospheric methane by tundra soils. Nature (London) 346:160-162.

Whalen, S.C., W.S. Reeburgh, and K.A. Sandbeck. 1990. Rapid methane oxidation in a landfill cover soil. Appl. Environ. Microbiol. 56:3405-3411.

Whittenbury, R., and N.R. Krieg. 1984. Methylococcaceae. p. 256-261. *In* N.R. Krieg and J.G. Holt (ed.) Bergey's manual of systematic bacteriology. Vol. 1. Williams & Wilkins, Baltimore, MD.

Williams, R.T., and R.L. Crawford. 1985. Methanogenic bacteria, including an acid-tolerant strain, from peatlands. Appl. Environ. Microbiol. 50:1542-1544.

Wolf, H.J., and R.S. Hanson. 1980. Identification of methane-utilizing yeasts. FEMS Microbiol. Lett. 7:177-179.

Wolin, M.J., and T.L. Miller. 1987. Bioconversion of organic carbon to CH_4 and CO_2. Geomicrobiol. J. 5:239-259.

Yoshinari, T. 1985. Nitrite and nitrous oxide production by *Methylosinus trichosporium*. Can. J. Microbiol. 31:139-144.

Zeikus, J.G. 1977. The biology of methanogenic bacteria. Bacteriol. Rev. 41:514-541.

11 Factors Affecting Methane Production in Flooded Rice Soils

Charles W. Lindau, William H. Patrick, Jr., and Ron D. DeLaune

Wetland Biogeochemistry Institute
Louisiana State University
Baton Rouge, Louisiana

The "greenhouse effect" is referred to as an increase in infrared absorption in the atmosphere, due to increasing concentrations of specific trace gases (Rowland, 1989). Concentrations of greenhouse gases have been steadily increasing in the atmosphere over the past 50 to 100 yr and may be contributing to an increase in global temperatures and O_3 depletion. Agricultural productivity, conversion of forestlands, burning of fossil fuels, industrial activity and natural wetlands have contributed to increases in atmospheric concentrations of carbon dioxide (CO_2), methane (CH_4), nitrous oxide (N_2O), water vapor (H_2O) and chlorofluoromethanes (NASA, 1988). Greenhouse gases are transparent to incoming solar radiation (ultraviolet) that reach the Earth's surface unhindered, but the trace gases are opaque and capable of absorbing the outflow of long-wave (infrared) radiation (Bouwman, 1990). Trace gas species can have significant effects on the thermal dynamics of the atmosphere if their long-wave absorption bands are in the partially opaque 8- to 12-μm region (Chamberlain et al., 1982). Methane has a strong absorption band at 7.66 μm, N_2O has infrared absorption bands at 7.8 and 17 μm, and CO_2 and chlorofluoromethanes have very strong absorption bands in the 8- to 12-μm region.

Methane is an important trace greenhouse gas that may account for about 15 to 20% of the total current increase to global warming. This may be caused by a 5- to 10-yr atmospheric residence time, but more importantly because 1 g of CH_4 will absorb about 70 times more infrared radiation than 1 g of CO_2 (USEPA, 1990). By 1978 the average tropospheric CH_4 concentration had increased to 1.5 ppmv and in 1988 increased to about 1.7 ppmv (Blake & Rowland, 1988). Globally, atmospheric CH_4 has increased about 1% yr^{-1} between 1965 and 1975. Trapped air data in dated ice cores suggest that atmospheric CH_4 concentrations have approximately doubled in the last 350 yr (Craig & Chou, 1982). It is estimated by the year 2100, atmospheric CH_4

concentrations may be about 4 ppmv or more than double over current levels (USEPA, 1990).

Estimates of total global CH_4 emissions from all sources range from 440 to 640 Tg yr^{-1} (Tg $= 10^{12}$ g) averaging about 540 Tg yr^{-1}. Anthropogenic sources are thought to account for approximately 60 to 70% of current CH_4 emissions (Cicerone & Oremland, 1988). Major human-related sources of CH_4 in teragrams per year include; rice (*Oryza sativa* L.) cultivation (60–170), animals (65–100), biomass burning (50–100), landfills (25–40), coal mining (30–50), and oil–gas systems produce 25 to 50 (Cicerone & Oremland, 1988; USEPA, 1990).

RICE CULTIVATION

Rice provides the major caloric source for 40 to 50% of the world's population and in less developed Asian countries 80 to 90% of the people depend on rice as a staple food (Vlek & Byrnes, 1986; Matthews et al., 1991). Asia accounts for approximately 90% of the total world rice area with the majority grown in the tropics and subtropics (De Datta, 1981). In 1985, approximately 146 $\times$ 10^6 ha of rice were harvested, of which 95% is located in the Far East and this harvested area accounts for 9.5% of the world's total cultivated area (FAO, 1985). Approximately 80% of the rice is grown under lowland or wetland conditions. In 1985, 440 million tons of rice were produced and based on projections of global population the demand for rice over the next 30 yr will increase by about 50%. By the year 2020 about 200 $\times$ 10^6 ha of rice will need to be grown to produce the 680 million tons of rice required by population increases (IRRI, 1988). Due to the projected increases in cultivated rice, CH_4 emissions over the next decade may increase by as much as 20% (USEPA, 1990).

In order to stabilize or reduce the atmospheric CH_4 concentration the biogeochemistry of methanogenesis and the factors governing CH_4 production in flooded rice systems need to be investigated. From this CH_4 research, field technologies and management practices must be developed to reduce CH_4 emissions without affecting flooded rice production.

FACTORS REGULATING METHANE PRODUCTION

Methanogenesis is an anaerobic microbial decomposition process of organic matter and any parameters affecting the chemical, physical, or biological characteristics of the flooded rice environment will influence the production of CH_4. Methanogens can function in almost any anaerobic ecosystem and can withstand salinity, pH, and temperature extremes, but can not function in the presence of O_2 and other oxidized inorganic compounds (Van Breemen & Feijtel, 1990). Factors influencing CH_4 production in flooded rice systems include: oxidation–reduction (redox) potential, pH, soil temperature, organic matter, soil type and addition of chemicals or fertilizers (USEPA, 1990).

Intensity of Soil Reduction (Redox Potential)

Flooded rice paddies are characterized by the absence of O_2 and activity of facultative and anaerobic bacteria. Within 6 to 8 h after submergence the soil is virtually O_2-free except at the floodwater–soil and rhizosphere–soil interfaces (Reddy & Patrick, 1984; Mikkelsen, 1987). Depletion of soil O_2 forces the facultative and obligate anaerobic microorganisms to use NO_3^-, Mn^{4+}, Fe^{3+}, SO_4^{2-} and CO_2 as electron acceptors for respiration and decomposition of organic matter (Van Breemen & Feijtel, 1990).

A well-oxidized soil has a redox potential of $+400$ to $+700$ mV but flooded rice soils may exhibit potentials as low as -250 to -300 mV (Patrick & DeLaune, 1977). Figure 11–1 shows the critical redox potential at which oxidized forms of inorganic substances are reduced and the sequential reduction must take place before CH_4 is generated. Oxygen is the first compound reduced after submergence at a redox potential of about $+350$ mV. Nitrate is then reduced to N_2O and N_2 at $+250$ mV. Manganic forms are reduced at about the same time or slightly later than NO_3^-. Ferric iron (Fe) is the next to get reduced and occurs at approximately $+125$ mV. When the redox potential drops to -150 mV, SO_4^{2-} is reduced to S^{2-} by specialized anaerobic bacteria (Connell & Patrick, 1968). After the disappearance of SO_4^{2-}, methanogens will start producing CH_4 at redox values of -200 to -250 mV if an energy source is available (Patrick & DeLaune, 1977; Bouwman, 1991). Compounds providing sources of C and energy for methanogenic bacteria are simple and major substrates include $H_2 + CO_2$, acetate, formate, methylated amines and methanol (Oremland, 1988; Schutz et al., 1989). Methanogens appear to be outcompeted by SO_4^{2-} reducers, and are gener-

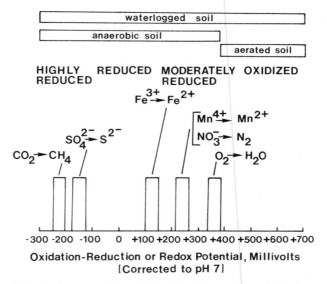

Fig. 11–1. Critical redox potential at which oxidized inorganic species begin to undergo reduction in submerged soils (Patrick & DeLaune, 1977).

Concentration(not to scale)

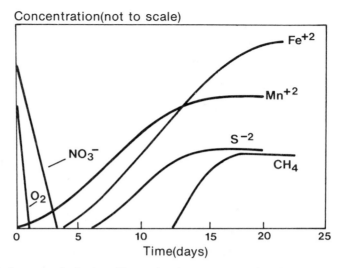

Fig. 11-2. Sequence of reduction of inorganic redox components in a soil after flooding (Patrick & Reddy, 1977).

ally inactive when SO_4^{2-} is present (Van Breemen & Feijtel, 1990). This low redox potential can be reached within 2 wk after submergence if the rice soil contains a sufficient energy source & low concentrations of NO_3^-, Mn^{4+} and Fe^{3+} (Fig. 11-2). The redox potential is critical in formation of CH_4 in rice paddies and high concentrations of oxidized inorganic forms may delay or inhibit methanogenesis. Field studies conducted in California, Louisiana, and Japan have shown that increasing CH_4 emissions are directly related to redox potential decreases in paddy soils (Cicerone et al., 1983; Yagi & Minami, 1990; Lindau et al., 1991).

Soil Temperature

Soil temperature plays an important role in soil microorganisms activity and CH_4 production in rice systems. In a laboratory bottle experiment, CH_4 production was measured in pure cultures of methanogenic bacteria incubated at different temperatures. Depending on the methanogen, the minimum, optimum and maximum temperatures for CH_4 formation were about 15, 35 and greater than 40 °C, respectively (Schutz et al., 1990). Holzapfel-Pschorn and Seiler (1986) reported that field CH_4 flux rates from an Italian rice paddy were strongly dependent on soil temperature (5-cm depth) and doubled as temperature increased from 20 to 25 °C. Methane fluxes correlated well with soil temperatures measured at the 1- to 10-cm depth but no significant correlation was observed at the 15-cm soil depth. In a detailed study, Schutz et al. (1990) investigated the influence of soil temperature on CH_4 fluxes from flooded rice fields. The data showed that diel changes in CH_4 emissions were significantly correlated to a particular soil temperature depth over the rice growing season. During May, June, and August, the best correlation occurred at the 1- to 5-cm depth and at the 10- to 15-cm soil depth during periods in June and July. The change in depth over the growing season may be a function of predominant CH_4 formation or CH_4 oxidation processes (Schutz et al., 1990).

Methane measurements from rice paddies in China also demonstrated · a strong relationship between flux and soil temperature measured at the 5-cm depth (Khalil & Rasmussen, 1991). Emissions of CH_4 from the Chinese rice paddies increased rapidly with increasing soil temperatures (18–31 °C) but the investigators noted the CH_4 flux increases were not due entirely to the temperature effect. In an earlier study, Cicerone et al. (1983) saw no clear correlation between soil temperature at the 10-cm depth and CH_4 flux rates from a California rice paddy.

Organic Matter

Addition of organic materials has been shown to dramatically increase CH_4 production from lowland rice fields. Schutz et al. (1989) reported addition of rice straw stubble to an Italian rice field increased CH_4 emissions. Rice straw application rates of 5 and 12 t ha^{-1} increased emission rates by factors of 2 and 2.4 over untreated control plots. Increasing application rates up to 60 t ha^{-1} did not increase CH_4 emission rates (Schutz et al., 1989).

Yagi and Minami (1990) determined the effects of added rice straw and rice straw compost on CH_4 evolution from Japanese rice paddies. Application of rice straw (6–9 t ha^{-1}) increased the CH_4 flux rates approximately 2 to 3.5 times compared to plots that did not receive rice straw. Addition of compost only slightly increased CH_4 emissions compared to control plots (Yagi & Minami, 1990). The large difference in CH_4 fluxes from rice straw and rice straw compost plots may be due to the C–N ratio of the organic matter applied. Composed organic material with lower C–N ratios may result in lower emissions of CH_4 (Yagi & Minami, 1990; USEPA, 1991).

Sass et al. (1991) also concluded that rice straw addition to Texas rice fields increased CH_4 emissions but noted grain yields dropped. Rice straw additions of 8 to 12 t ha^{-1} increased CH_4 emissions about 2 to 2.5 times compared to plots receiving urea–N but no rice straw. The decrease in grain yield was thought to be due to excessive anaerobiosis caused by the rice straw (Sass et al., 1991).

In addition to incorporated rice straw or compost, it has been suggested that the growing rice plant may also enhance CH_4 production from rice paddies. Field research indicates that during certain rice growth stages the roots may provide increasing amounts of root exudates and litter that can be used as additional substrates by methanogens to increase CH_4 fluxes. The emission periods appear to occur around the reproductive and rice heading stages (Schutz et al., 1989; Lindau et al., 1991; Sass et al., 1991).

Fertilizer

The relationship between fertilizer application and CH_4 emissions from flooded rice systems is not clear. Methane fluxes are strongly influenced by type, method of application, and rate of fertilizer application. Cicerone & Shetter (1981) reported $(NH_4)_2SO_4$ addition (140 kg N ha^{-1}) to California rice plots increased CH_4 fluxes by about fivefold compared to unfertilized control plots. Schutz et al. (1989) evaluated fertilizer applications to Italian rice paddies over a 3-yr cropping period. Addition of urea–N (100 and 200 kg N ha^{-1}) reduced CH_4 emissions by approximately 18% compared to control plots but was strongly influenced by mode of application. Methane emissions increased as the mode of urea–N application changed; surface applied > raked into soil (5-cm depth) > incorporated (20-cm depth). Deep application of urea resulted in a 40% decrease in CH_4 emissions compared to shallow urea incorporation. Surface applied urea–N enhanced total CH_4 fluxes by about 19% compared with unfertilized control plots (Schutz et al., 1989). Methane emission rates from $(NH_4)_2SO_4$-treated plots were much lower compared to unfertilized Italian control plots and independent of application rates (50–200 kg N ha^{-1}). On a seasonal average CH_4 fluxes were lower by 6, 43 and 62% when $(NH_4)_2SO_4$ was surface applied, raked and incorporated into the soil, respectively (Schutz et al., 1989). Lindau et al. (1991) reported that increasing amounts of urea–N fertilizer applied to flooded Louisiana rice plots had a significant effect on CH_4 emissions. Urea-nitrogen was surface applied, preflood at rates of 0, 100, 200, and 300 kg N ha^{-1} and CH_4 evolved over the 86-d sampling period was about 210, 300, 310, and 370 kg ha^{-1}, respectively. The CH_4 emission increases were thought to be due to increased plant growth, increased concentrations of root exudates and greater root growth (Lindau et al., 1991).

The processes that reduce CH_4 emissions after $(NH_4)_2SO_4$ additions (in some reported field experiments) are not fully understood. Methane formation would be expected to be inhibited by the presence of SO_4^{2-} due to the order of reduction of oxidized inorganic compounds (Patrick & DeLaune, 1977). Sulfate-reducing bacteria may also be outcompeting methanogenic bacteria for substrates and certain soil SO_4^{2-} and S^{2-} concentrations may be toxic to methanogens (Jakobsen et al., 1981, Kristjansson et al., 1982). It has also been suggested that reoxidation of S^{2-} to SO_4^{2-} in the rice rhizosphere may maintain CH_4 inhibition over longer periods of time in the flooded rice fields (Freney et al., 1982).

Soil Types

Methane emission rates vary widely with soil types. Yagi and Minami (1990) measured CH_4 fluxes from four Japanese paddy soils and reported that CH_4 evolution rates differed widely with soil type. Annual CH_4 emission rates from the Japanese paddy soils, located in the same climatic region, were; peaty soils > alluvial soil > Andosol. Annual CH_4 flux rates from the peat soil was about 40 times higher than the Andosol soil. In addition, the percolation rate in the Andosol soils was three times the peat soil that may contribute to the lower CH_4 emissions observed (Yagi & Minami, 1990). Sass et al. (1991) examined two Texas rice soils, located near Beaumont, TX on CH_4 emissions over a two-yr period. Methane emissions from the Lake Charles clay (Typic Pelludert) site were about three times greater than measured fluxes from the Beaumont clay (Entic Pelludert) rice field. The large difference in flux rates may have been due to differences in soil compactness (Sass et al., 1991). Additional studies in India have also shown that soil type significantly affects CH_4 evolution. Four Indian paddy soils with pH's ranging from 7.0 to 8.8 showed that the optimum pH range for methanogenesis was about 7.5 to 8.5 and it was also noted that soil texture played a significant role in CH_4 production (Parashar et al., 1991).

Soil pH

Methane production in flooded rice systems is favored by a near neutral soil pH (7) but is influenced by soil type (Bouwman, 1991). Jakobsen et al., (1981) determined that the optimum pH for CH_4 formation in a Crowley silt loam rice soil suspension was 6.7. Most methanogenic bacteria grow over a narrow range of pH from about 6 to 8 (Jenkins, 1963; Oremland, 1988).

SUMMARY

Methane is an important greenhouse gas and may account for 15 to 20% of the current increase in commitment to global warming and tropospheric concentrations are increasing about 1% yr^{-1}. Flooded rice paddies are a major source of CH_4 and contribute 60 to 170 Tg to the atmosphere per year that is approximately 25% of total global CH_4 emissions. To keep up with population increases over the next 20 to 30 yr it is estimated that about 50 $\times$ 10^6 ha of land will have to be converted to flooded rice production, which may increase CH_4 emissions by 20% in the next decade.

Flooded systems are conducive to the production of CH_4 due to the presence of methanogenic bacteria that decompose organic materials under anaerobic soil conditions. Methane production in flooded rice soils is strongly influenced by soil factors such as; redox potential, temperature, pH, organic matter, soil type and addition of chemicals or fertilizers. An understanding of the factors and related processes regulating CH_4 production is needed in order to stabilize or reduce future CH_4 emissions from flooded rice fields. Such information hopefully will lead to field management practices that will reduce CH_4 emissions while maintaining or increasing the productivity of lowland rice.

ACKNOWLEDGMENTS

This research was supported in part by the Department of Energy (Award no. DE-FC03-90ER61010).

REFERENCES

Blake, D.R., and F.S. Rowland. 1988. Continuing worldwide increase in tropospheric methane, 1978 to 1987. Science (Washington, DC) 239:1129-1131.

Bouwman, A.F. 1990. Background. p. 25-192. *In* A.F. Bouwman (ed.) Soils and the greenhouse effect. John Wiley and Sons, New York.

Bouwman, A.F. 1991. Agronomic aspects of wetland rice cultivation and associated methane emissions. Biogeochemistry 15:65-88.

Chamberlain, J.W., H.M. Folex, G.J. MacDonald, and M.A. Ruderman. 1982. Climate effects of minor atmospheric constituents. p. 253-278. *In* W.C. Clark (ed.) Carbon dioxide review. Oxford Univ. Press, New York.

Cicerone, R.J., and J.D. Shetter. 1981. Sources of atmospheric methane: Measurements in rice paddies and a discussion. J. Geophy. Res. 86:7203-7209.

Cicerone, R.J., J.D. Shetter, and C.C. Delwiche. 1983. Seasonal variations of methane flux from a California rice paddy. J. Geophys. Res. 88:11 022-11 024.

Cicerone, R.J., and R.S. Oremland. 1988. Biogeochemical aspects of atmospheric methane. Global Biogeochem. Cycles 2:299-327.

Connell, W.E., and W.H. Patrick, Jr. 1968. Sulfate reduction in soil: Effects of redox potential and pH. Science (Washington, DC) 159:86-87.

Craig, H., and C.C. Chou. 1982. Methane: The record in polar ice cores. Geophys. Res. Lett. 9:1221-1224.

DeDatta, S.K. 1991. Principles and practices of rice production. John Wiley and Sons, New York.

Food and Agriculture Organization of the United Nations. 1985. Fertilizer yearbook. Vol. 35. FAO Statistical Ser. no. 71. FAO, Rome.

Freney, J.R., V.A. Jacq, and J.F. Baldensperger. 1982. The significance of the biological sulfur cycle in rice production. Dev. Plant Soil Sci. 5:271-317.

Holzapfel-Pschorn, A., and W. Seiler. 1986. Methane emission during a cultivation period from a Italian rice paddy. J. Geophys. Res. 91:11 803-11 804.

International Rice Research Institute. 1988. World rice statistics. 1987. IRRI, Los Baños, Philippines.

Jakobsen, P., W.H. Patrick, Jr., and B.G. Williams. 1981. Sulfide and methane formation in soils and sediments. Soil Sci. 132:279-287.

Jenkins, D. 1963. Sewage treatment p. 508-536. *In* C. Rainbow and A.H. Rose (ed) Biochemistry of industrial microorganisms. Academic Press, New York.

Khalil, M.A.K., and R.A. Rasmussen. 1991. Methane emissions from rice fields in China. Environ. Sci. Technol. 25:979-981.

Kristjansson, J.K., P. Schonheit, and R.K. Thauer. 1982. Different Ks values for hydrogen and methanogenic and sulfate-reducing bacteria: An explanation for the apparent inhibition of methanogenesis by sulfate. Arch. Microbiol. 131:278-282.

Lindau, C.W., P.K. Bollich, R.D. DeLaune, W.H. Patrick, Jr., and V.J. Law. 1991. Effect of urea fertilizer and environmental factors on CH_4 emissions from a Louisiana USA, rice field. Plant Soil 136:195-203.

Matthews, E., I. Fung, and J. Lerner. 1991. Methane emission from rice cultization: Geographic and seasonal distribution of cultivated areas and emissions. Global Biogeochem. Cycles 5:3-24.

Mikkelsen, D.S. 1987. Nitrogen budgets in flooded soils used for rice production. Plant Soil 100:71-97.

Minami, K. 1989. Effects of agricultural management on methane emissions from rice paddies. *In* Proc. of the Workshop on Greenhouse Gas Emissions from Agricultural Systems of the IPCC Response Strategies Working Group, 12-14 December. USEPA/USDA, Washington, DC.

National Aeronautics and Space Administration. 1988. Earth system science, a closer view. NASA, Washington, DC.

Oremland, R.S. 1988. Biogeochemistry of methanogenic bacteria. p. 641-705. *In* A.J.B. Zehnder (ed.) Biology of anaerobic microorganisms. John Wiley and Sons., New York.

Parashar, D.C., J. Rai, K. Gupta, and N. Singh. 1991. Parameters affecting methane emission from paddy fields. Ind. J. Radio Space Phys. 20:12-17.

Patrick, W.H., Jr., and R.D. DeLaune. 1977. Chemical and biological redox systems affecting nutrient availability in the coastal wetlands. Geosci. Man 18:131-137.

Patrick, W.H., Jr., and C.N. Reddy. 1977. Chemical changes in rice soils. p. 361–379. *In* IRRI symp. on soils and rice. IRRI, Los Bano., Philippines.

Reddy, K.R., and W.H. Patrick, Jr. 1984. Nitrogen transformations and loss in flooded soils and sediments. p. 274–309. *In* C.P. Straub (ed.) CRC Crit. Rev. Environ. Control. Vol. 13.

Rowland, F.S. 1989. Chlorofluorocarbons and the depletion of stratospheric ozone. Am. Sci. 77:36–45.

Sass, R.L., F.M. Fisher, and P.A. Harcombe. 1991. Mitigation of methane emissions from rice fields: Possible adverse effects of incorporated rice straw. Global Biogeochem. Cycles 5:275–287.

Schutz, H., A. Holzapfel-Pschorn, R. Conrad, H. Rennenberg, and W. Seiler. 1989. A three year continuous record on the influence of daytime, season, and fertilizer treatment on methane emission rates from an Italian rice paddy. J. Geophys. Res. 94:16 405–16 416.

Schutz, H., W. Seiler, and R. Conrad. 1990. Influence of soil temperature on methane emission from rice paddy fields. Biogeochemistry 11:77–95.

U.S. Environmental Protection Agency. 1990. Methane emissions and opportunities for control. EPA/400/9-90/007. USEPA, Washington, DC.

U.S. Environmental Protection Agency. 1991. Introduction: Methane, global warming, and wetland rice. p. 12–18. *In* B.V. Braatz and K.B. Hogan (ed.) Sustainable rice productivity and methane reduction research plan. USEPA, Office of Air and Radiation, Washington, DC.

Van Breemen, N., and T.C.J. Feijtel. 1990. Soil processes and properties involved in the production of greenhouse gases, with special relevance to soil taxonomic systems. p. 195–223. *In* A.F. Bouwman (ed.). Soils and the greenhouse effect. John Wiley and Sons, New York.

Vlek, P.L.G., and B.H. Byrnes. 1986. The efficacy and loss of fertilizer N in lowland rice. Fert. Res. 9:131–147.

Yagi, K., and K. Minami. 1990. Effect of organic matter application on methane emission from some Japanese paddy soils. Soil Sci. Plant Nutr. 36:599–610.

12 Controls on Methane Flux from Terrestrial Ecosystems

Joshua P. Schimel

Institute of Arctic Biology
University of Alaska
Fairbanks, Alaska

Elisabeth A. Holland

National Center for Atmospheric Research
Boulder, Colorado

David Valentine

Natural Resources Ecology Laboratory
Colorado State University
Fort Collins, Colorado

Research on CH_4 dynamics has been stimulated in recent years by the finding that atmospheric CH_4 has been increasing at approximately 1% yr^{-1} (4 times as fast as CO_2; Bouwman, 1989) and that it is approximately 30 times as effective a greenhouse gas as CO_2 (Bouwman, 1989). Methane has other critical roles in atmospheric chemistry: it reacts with the OH radical, a key species in atmospheric oxidation chemistry; and CH_4 oxidation is a major source of stratospheric H_2O vapor as well (Schütz et al., 1991). The importance of atmospheric CH_4 and its increase has challenged our understanding of the CH_4 cycle and posed the following questions: (i) What are CH_4 fluxes to and from the atmosphere?, (ii) How have the fluxes changed since preindustrial times?, and (iii) How will fluxes change over the next several decades?

What might be called the first phase of global CH_4 research has focused on tightening estimates of CH_4 sources and sinks, with a fair degree of success (Cicerone & Oremland, 1988; Khalil & Rasmussen, 1990; Fung et al., 1991). While some uncertainties still exist, it has become clear that net CH_4 fluxes from soils (predominantly wetlands) are approximately 180 to 220 Tg yr^{-1} and that they comprise about 40 to 43% of the total global source (Cicerone & Oremland, 1988; Fung et al., 1991). Soil sources are dominated by rice (*Oryza sativa*) paddies (100 Tg) and natural wetlands (115 Tg, of which

 Agricultural Ecosystem Effects on Trace Gases and Global Climate Change. ASA Special Publication no. 55.

35 Tg is from high latitude sources). Estimates of CH_4 oxidation are more varied, ranging from 10 to 50 Tg or 2 to 15% of the atmospheric oxidation by reaction with the OH radical (450–500 Tg; Born et al., 1990; Fung et al., 1991).

Addressing the changes in fluxes is much more difficult and needs to be the focus of future CH_4 research, with the ultimate question: How will CH_4 fluxes change under conditions of altered climate? To answer this, we must improve our understanding of the ecosystem-level controls on CH_4 dynamics. This will enable us to develop large-scale models of CH_4 cycling, which can provide estimations and extrapolations of CH_4 fluxes over space and time. Soil scientists have an invaluable contribution to make in this phase of CH_4 research, as the soil processes are the most complex in the CH_4 cycle and they are very sensitive to climate.

In this paper, we will identify some of the key controls on CH_4 flux between terrestrial ecosystems and the atmosphere, and discuss new research that begins to address them. This is not intended as a general discussion of all current research on terrestrial CH_4 dynamics, but rather a discussion of what we feel to be particularly pressing issues in being able to model and predict CH_4 fluxes in terrestrial ecosystems.

In considering the processes that control the terrestrial CH_4 flux to the atmosphere, we also need to consider the different types of ecosystems in which they occur. These can be divided into wetland and nonwetland systems (hereafter referred to simply as upland systems). Wetland systems are flooded for a period of the year, and have the capacity to produce substantial amounts of CH_4 on a regular basis; these systems may also have highly elevated CH_4 concentrations in the soil (Brown et al., 1989). Upland systems rarely produce CH_4 (only during episodic saturated events), their only role in the CH_4 cycle is as a sink for atmospheric CH_4, and CH_4 concentrations higher than the atmospheric 1.7 ppm are rare.

Methane flux from any ecosystem is controlled by several processes:

$$\text{Flux} = \text{production} - \text{consumption} \pm \text{storage}.$$

Each of these terms, production, consumption, and storage, provide important foci of current and future research. An effective model of CH_4 emissions must incorporate each of these terms as well as CH_4 transport through soil, water, and plants to the atmosphere. In this chapter, we focus on CH_4 production, CH_4 consumption, their interactions and the role of plant and soil-mediated transport to the atmosphere. Methane storage in soils and its role in CH_4 emissions has received relatively little attention (Brown et al., 1989) and will not be discussed further.

METHANE EFFLUX

Terrestrial wetlands, including extensive peatlands, contribute approximately 15 to 25% of the global CH_4 sources (Schütz et al., 1991). Peatlands

Processes controlling CH₄ flux

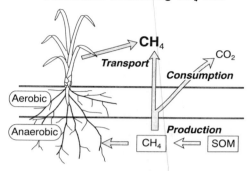

Fig. 12–1. Processes controlling CH_4 flux from wetland soils.

are generally saturated for at least part of the year, producing a vertical stratification with aerobic zones, where CH_4 is consumed, overlying anaerobic zones, where CH_4 is produced. For CH_4 to be released to the atmosphere, it must be transported through the aerobic surface zone without being oxidized. Thus the balance between production and consumption becomes the overriding control on actual CH_4 efflux from the soil system and an effective model of CH_4 emissions will include explicit consideration of both of these processes. This poses three questions for soil scientists: (i) What controls the total production of CH_4?, (ii) What is the fate of CH_4?, and (iii) What controls that fate?

Central to both of these issues is the activities of plants (Fig. 12–1). Plants supply C to the soil that is converted to CH_4. Many wetland plants have aerenchyma that can transport CH_4 through the root system to the atmosphere, bypassing CH_4 consumption in the soil of the surface zone (Mariko et al., 1991; Schütz et al., 1991).

Controls on Methanogenesis in Wetlands

Methanogenesis, the biological production of CH_4, occurs only under anaerobic conditions when all electron acceptors other than CO_2 or acetate have been reduced (e.g., O_2, NO_3^-, Fe^{3+}, SO_4^{2-}). Thus, methanogenesis occurs only in ecosystems that are chronically flooded and with sufficient levels of microbial activity to utilize O_2 present (Conrad, 1989).

Given appropriate redox conditions, C substrate availability is the major proximal, or physiological, control on CH_4 production (Fig. 12–2). Secondary controls on methanogenesis are pH and temperature; temperature, however, probably primarily regulates the supply of substrate to methanogens. Plants are critical in supplying C to methanogens. In natural wetlands peat is recalcitrant and in rice paddies rapid decomposition often produces low organic matter levels. Fresh root exudates and residues are therefore a critical source of labile C for methanogenesis (Schütz et al., 1991). This is further supported by work of Asselmann and Crutzen (1988) suggesting that

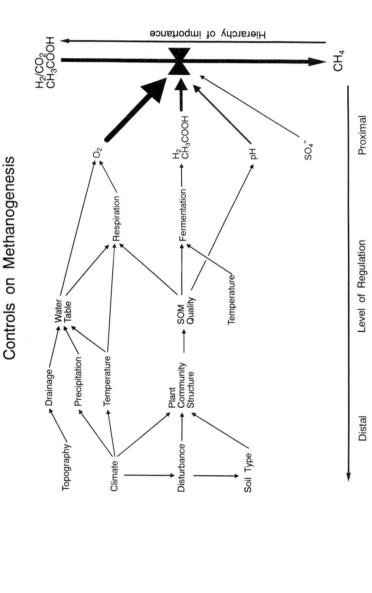

Fig. 12-2. Relationships between proximal and distal controls on CH_4 production in anaerobic soils.

net primary production in an ecosystem is predictive of CH_4 production rates.

Our research has attempted to consider each of these controls in two ecosystems characterizing endpoints in a successional sequence: a young minerotrophic fen (< 1000 yr old) located near James Bay, ON, called the "Coastal Fen" and older ombrotrophic bog (approximately 5000 yr old) located approximately 100 km inland, called "Kinosheo Lake." We developed quantitative relationships between methanogenesis and C availability, temperature and pH (Fig. 12–3).

Carbon Availability

One difficulty in developing relationships between C availability and CH_4 production is that most C substrates must be fermented to either acetate or to H_2 before being converted to CH_4, with acetogenic pathways accounting for more than 70% of CH_4 production in terestrial ecosystems where oceanic influences are minimal (Knowles, 1992).

We have been carrying out studies to develop quantitative models of CH_4 production in peat soils. Very low rates of soil respiration together with low rates of CH_4 production suggest that very little of the C present in these peats was available for microbial processing (Valentine et al., 1992). Kinosheo Lake peats produced very little CH_4 except in "Black Holes" that are pockets of H_2O and degraded peat about 2 to 3 m in diameter. Black Holes cover a small proportion of the total surface area but account for most of the total CH_4 flux (N. Roulet, 1991, personal communication). Rates of CH_4 production were considerably lower in Black Hole than in Coastal Fen peats. The Kinosheo Lake site had deep peat deposits characterized by low pH that may have further limited net primary production and slowed C turnover (L. Klinger, 1992, personal communication).

In month-long anaerobic slurries, C additions resulted in significant increases in rates of methanogenesis. To accelerate the production of acetate and other labile C precursors, we added 16.44 mmol of C as ethanol to the slurries (82.2 mM ETOH). This nearly doubled rates of CH_4 production from Coastal Fen peats. Similarly, additions of similar amounts of labile C (assuming that 25% of litter C was labile) as ground plant material resulted in a 10-fold increase in CH_4 production rates, though this was very variable. The EtOH additions had an even greater effect on CH_4 production rates in the Black Hole samples where EtOH additions increased CH_4 production five- to six-fold.

Other Controls

Temperature and pH also provided important controls over CH_4 production. Varying temperatures from 10 to 20 °C increased CH_4 production rates three- to four-fold from Coastal Fen peats, but had little effect on Kinosheo Lake peats. The temperature control, however, is probably exerted on the rate of fermentation of organic C to acetate and other precursors, rather than on methanogenesis itself (Conrad, 1989).

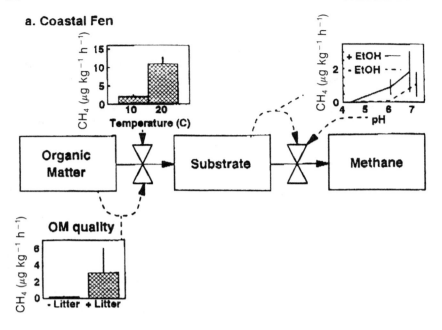

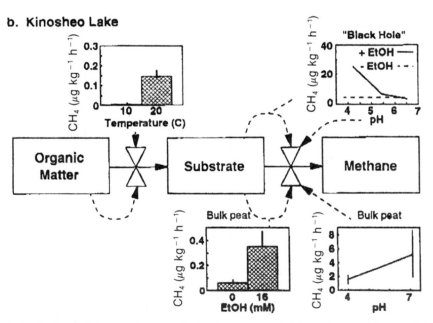

Fig. 12-3. Constraints on methane production in peats sampled from a) Coastal Fen, and b) Kinosheo lake, "+EtOH" treatments refers to addition of 16.4 mmol C as ethanol to 100 mL slurries, yielding 82.2 mM EtOH.

Methanogenesis in vitro is generally most rapid under near neutral or slightly alkaline pH conditions (Knowles, 1992; Yarrington & Wynn-Williams, 1985). Our results generally support this conslusion; we found that lowering the pH of Coastal Fen incubations from ambient ($\sim$7) to 5.5 or 4 dramatically reduced or eliminated methanogenesis, both with and without ethanol. In Kinosheo Lake peat, with a natural pH of 3.9, NaOH additions that raised the pH to 6.2 resulted in a threefold increase in already low rates of CH_4 production. The exception to this pattern was the Black Hole samples, in which increasing pH did not affect CH_4 production in unamended peat and negated the stimulation by EtOH additions.

Our data suggest that the primary constraint on CH_4 production is substrate availability, modified further by temperature and pH constraints. This is consistent with the view that net primary production is a good predictor of CH_4 production rates (Asselmann & Crutzen, 1988). Methane production rates in laboratory incubations, however, do not translate directly into field rates of CH_4 production. A major challenge for future research is to apply relationships gleaned in the laboratory to field estimates.

Fate of Methane in Wetland Systems

Methane is produced at depth in wetland soils, and must be transported through the soil to reach the atmosphere. During the process of transport, CH_4 has the potential to be oxidized in the surface aerobic layer (Whalen et al., 1990; Yavitt et al., 1990) or in the rhizosphere of aerenchymous plants, which transport O_2 into the soil. Estimates of the total amount of CH_4 consumed in wetlands before it can be released to the atmosphere range from 40 to 95% of the total CH_4 produced (Galchenko et al., 1989).

Plants also have the potential to transport substantial quantities of CH_4 (Schütz et al., 1991). This is highlighted by data we collected from arctic wet meadow tundra (Table 12–1), in which we measured total CH_4 flux

Table 12–1. Importance of CH_4 consumption and plant transport in wet meadow tundra.†

Community	Fluxes			
	Total flux	Bare soil	Estimated plant transport‡	% Plant transport
		mg m^{-2} d^{-1}		
Eriophorum angustifolium	200.4	−1.3	125.7	63–100
Carex aquatilis	14.0	0.1	7.7	55–99
Mixed§	86.8	0.8	88.1	99–100
Mixed	101.0	0.5	82.4	81–99

† Study site wa on the shore of Toolik Lake, AK, adjacent to the Arctic Tundra LTER site.
‡ Flux through individual plants was measured by sealing a tiller into a chamber (made from a 1-L nalgene bottle) with a split rubber stopper that had a hole drilled into it; the hole was large enough to avoid damaging the plant. Total plant flux was estimated by multiplying the average flux for each species by the number of tillers in the main chamber. *Eriophorum scheuchzeri* was not measured directly but was assumed to transport the same amount of CH_4 per unit tiller area as *E. angustifolium*.
§ Contained *E. angustifolium*, *E. scheuchzeri*, and *C. aquatilis*.

from the community using a static chamber (Hutchinson & Livingston, 1993), flux from bare soil adjacent to the chamber, and flux through individual plant tillers of species found in the chamber. Clearly, the bulk of the CH_4 escaping the soil was transported through plants, and essentially no CH_4 appeared to be escaping from bare soil. In fact, in one case the soil consumed CH_4, even while CH_4 was escaping through the plants. In other systems, particularly those with standing water, ebullition (bubbles) can be a major mechanism of CH_4 flux as well (Conrad, 1989).

The balance between CH_4 consumption and CH_4 transport to the atmosphere is central to controlling CH_4 flux, and understanding this balance is essential for modeling and predicting CH_4 dynamics in wetland. Modeling CH_4 transport will require collaboration with soil physicists, to incorporate diffusion and ebullition, and with plant physiologists to incorporate plant transport, a process that is still poorly understood (Schütz et al., 1991). Methane consumption is the other critical process controlling the fate of CH_4, and its extent and control also are poorly understood. Several important controls on CH_4 consumption have been identified, but others may still be unknown, and models of CH_4 consumption are lacking. The known controls on CH_4 consumption will be discussed in the next section, combined with discussion of CH_4 consumption in upland systems.

METHANE CONSUMPTION

Role of Methane Consumption

Methane consumption occurs in all aerobic soils and can serve two distinct ecological roles. First, CH_4 consumption in the aerobic surface soil and rhizosphere of wetlands can provide a barrier to efflux of CH_4 produced at depth, and if the water table drops far enough it it can become a sink (Harriss et al., 1982). Second, CH_4 consumption in upland soils acts as a net sink for atmospheric CH_4. These issues raise several important questions: (i) What are the magnitudes of the net and gross soil CH_4 sinks?, and (ii) What controls CH_4 consumption in soil?

Current estimates of net CH_4 consumption range from about 10 to 50 Tg yr^{-1}, which comprises between 2 and 15% of the atmospheric sink for CH_4 (Fung et al., 1991). Estimates of CH_4 consumption in upland soils have been very limited, with studies in temperate forests (Born et al., 1990; Steudler et al., 1990; Yavitt et al., 1990), Boreal forests (Whalen et al., 1991), grasslands (Seiler et al., 1984; Mosier et al., 1991), and tropical forests (Keller et al., 1983, 1986, 1990). The direct measurements suggest that the actual soil sink is probably somewhat greater than the 10 Tg yr^{-1} suggested by the latest estimate from a three-dimensional atmospheric model (Fung et al., 1991). It is important to resolve this disagreement and to tighten the estimate of the soil sink; this will require more measurements, particularly in biomes and land-uses that have not yet been studied.

Controls on Methane Consumption

Predicting changes in CH_4 over time requires incorporating the major controls on CH_4 consumption (oxidation) into large-scale models. Developing such models requires considering the proximal controls, i.e., those that directly affect the CH_4 consumers. There are two types of organisms that may be important CH_4 consumers in nature; methanotrophs and nitrifiers (Bédard and Knowles, 1989). Perhaps conveniently, the controls on CH_4 consumption by both groups of organisms appear to be similar (Bédard and Knowles, 1989). As we don't know the relative importance of nitrifiers vs. methanotrophs, we won't distinguish further, but this remains an important question.

We might expect somewhat different controls to apply between wetland and upland populations even at the physiological level, because methanotrophs in wetlands may regularly be exposed to extremely high CH_4 concentrations, while upland populations should be selected for survival at atmospheric CH_4 concentrations. Nitrifiers also should be uncommon in natural wetlands and peatlands because these systems are often acidic, NH_4^+-poor, and low in O_2, all characteristics selecting against nitrifiers (Schmidt, 1982). The significance of nitrifiers in CH_4 consumption in upland systems and rice fields could vary widely.

In nature, several factors appear to control CH_4 consumption rates, with the CH_4 supply to the organisms oxidizing the CH_4 as the most critical (Figs. 12–4, 12–5). Nitrogen availability appears to be an important control on CH_4 consumption in upland systems (Fig. 12–4), while O_2 may be important in wetlands (Fig. 12–5).

Methane Supply

Methane supply to the methanotrophs is likely to be the single greatest control on CH_4 consumption rates in all soils, wetland and upland. In wetland soils, consumption can account for up to 100% of the CH_4 diffusing up through the aerobic surface layer (Table 12–1). Thus total consumption is probably controlled dominantly by the rates of production and upward diffusion (Fig. 12–5).

In upland soils the low atmospheric CH_4 concentration and obstacles to gas diffusion should limit CH_4 supply to consumers (Fig. 12–4). Given current atmospheric CH_4 concentrations, CH_4 solubility in the soil solution is approximately 2.5 nM (Conrad, 1984). The lowest recorded K_m values (half-saturation constant) for CH_4 consumption are between 1 and 2 μM (Bédard & Knowles, 1989; Whalen et al., 1990). Methane concentrations in soil are therefore much lower than K_m values measured for CH_4 consumption, suggesting that methanotrophs in soil are strongly CH_4 limited. Methane consumption in upland soils generally shows a first-order concentration response (Yavitt et al., 1990; Whalen et al., 1991), which is what would be expected for processes occurring at concentrations below the organisms K_m.

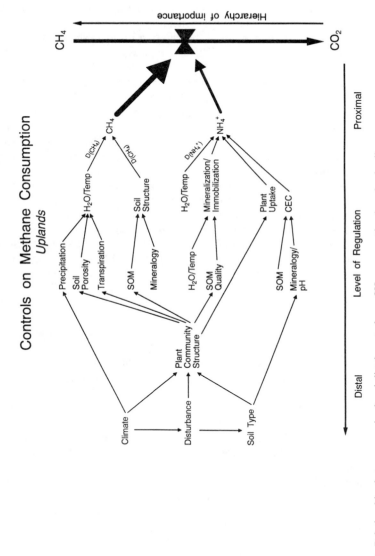

Fig. 12–4. Relationships between proximal and distal controls on CH_4 consumption in upland soils.

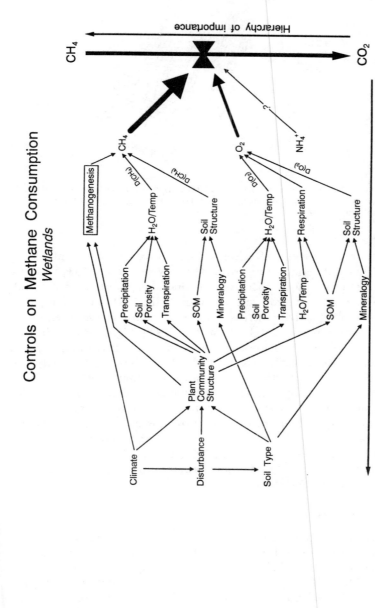

Fig. 12-5. Relationships between proximal and distal controls on CH_4 consumption in wetland soils.

Methane Flux in Tussock Soil

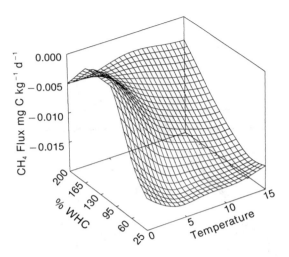

Fig. 12–6. Effect of temperature and moisture on CH_4 consumption in a tussock–tundra soil. Hand-mixed soils were incubated in sealed jars at defined moisture content (% of water holding capacity) and temperature.

The importance of gas diffusion is documented by Born et al. (1990) who found that soil permeability correlated very well to net CH_4 flux in a range of German forest soils. In a given soil, increasing water content usually decreases CH_4 consumption rates (Mosier et al., 1991, Fig. 12–6), even while total microbial activity increases. Methane consumption often shows little temperature response as well (Born et al., 1990; Fig. 12–6). In the tundra soil used for Fig. 12–6, microbial respiration increased substantially with increasing temperature (Kielland & Schimel, 1991). These data strongly suggest physical rather than biotic controls on CH_4 consumption. A last piece of evidence supporting physical rather than biological control is the finding that maximum rates of CH_4 consumption across a diverse array of ecosystems are frequently in the range of 3 to 4 mg m^{-2} d^{-1}, with average rates in the range of 1 to 2 mg m^{-2} d^{-1} (Born et al., 1990; Steudler et al., 1990; Yavitt et al., 1990; Mosier et al., 1991; Fig. 12–7). If CH_4 consumption was under predominantly biotic controls, one might expect a wider range of maximum and average rates.

Nitrogen Availability

Nitrifiers and methanotrophs can oxidize both CH_4 and NH_4^+, due to the similarity in shape and size of these substrates (tetrahedra with 31.1 vs. 29.7 nm Van der Waals radii for CH_4 and NH_3, respectively; Weast, 1976, p. D–178) and the relatively low specificity of the monooxygenase enzymes responsible (Bédard & Knowles, 1989). The ability to use both substrates sets

Methane Consumption in an Alaskan Birch Forest

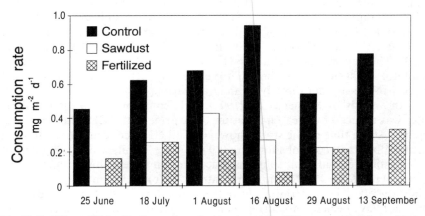

Fig. 12-7. Effect of N fertilization and sawdust treatments on CH_4 consumption in an Alaskan birch forest stand. Fertilizer plots had $(NH_4)_2SO_4$ added annually—sawdust plots had enough sawdust added to increase the C–N ratio of the forest floor to 50. Fluxes were measured using static chambers on permanently placed bases (Whalen et al., 1991).

up the possibility for competitive inhibition of CH_4 consumption by NH_4^+ (Fig. 12–4).

Several field studies over the last few years have shown strong inhibition of CH_4 consumption by NH_4^+-fertilizer additions. Nitrogen fertilization reduced CH_4 consumption in red pine (*Pinus resinosa* Ait) and mixed hardwood stands by up to 33% (Steudler et al., 1990) and in a Colorado pasture by 41% (Mosier et al., 1991). Mosier et al. (1991) also found that high fertility swale sites had generally lower rates of CH_4 consumption than a lower fertility midslope site and the swale sites showed no fertilization effect, unlike the midslope sites. These data suggest that CH_4 consumption is generally suppressed by rapid N turnover.

In Alaskan boreal forests, the effects of N fertilization is somewhat mixed. Whalen et al. (1991) reported no effect of N fertilization on CH_4 consumption in a range of communities. During the summer of 1991, however, we found significant inhibition of CH_4 consumption in an upland birch (*Betula papyrifera*) forest (Fig. 12–7). In this site, a sawdust-amended plot also had significantly lowered CH_4 consumption rates. This was particularly interesting as sawdust increases the C–N ratio of the forest floor and should immobilize N, in effect it acts as an antifertilizer. The sawdust may have acted as a mulch and retained moisture in the forest floor, reducing CH_4 diffusion, but the mechanism of inhibition is still unclear.

The field data then, appear to support the idea that high NH_4^+ availability inhibits CH_4 consumption. We feel, however, that the actual mechanism may not be purely enzymatic competition. If the inhibition is a purely enzymatic competition, then the controlling factor must be the concentration of NH_4^+ in the cells of the nitrifiers or methanotrophs carrying out the oxidation. The inhibition of CH_4 oxidation by NH_4^+ in methanotrophs has a measured K_i (NH_4^+ concentration in the growth medium at one-half maximum inhibition) of 10 mM at pH $\leq$ 7, which is roughly three orders of magnitude higher than the K_m values for NH_4^+ uptake and assimilation in bacteria (Bédard & Knowles, 1989; Button, 1985; Rosenfeld & Brenchley, 1983). This suggests that cellular NH_4^+ concentrations may only rarely reach inhibitory concentrations, though it is extremely difficult to relate this kind of physiological data directly to field activities.

In a study of Mosier et al. (1991), CH_4 consumption in the midslope sites was reduced by urea fertilization 10 yr prior to the CH_4 flux measurements. At the time of the study, however, soil NH_4^+ concentrations in both the fertilized and control sites were 0.9 mg N kg^{-1} soil, which is equivalent to an estimated bulk soil solution concentration of less than 1 mM, which is below the K_i (half inhibition concentration) for NH_4^+ on CH_4 oxidation. Since the NH_4^+ concentrations were the same, it suggests that the reduction may not have been due to current competitive inhibition. If the mechanisms of inhibition of CH_4 oxidation is not direct enzymatic competition, it remains unclear what it is. Verifying the mechanism of inhibition is a major issue that needs to be resolved if we are to quantify and model the relationship between CH_4 and N dynamics.

The role of N cycling in controlling CH_4 consumption in wetlands has been even less studied than in upland systems. Field studies in rice paddies show greater CH_4 flux with increasing urea fertilization (Lindau et al., 1991), but this was attributed primarily to increased productivity and C availability, rather than reduced consumption, though this could have been occurring as well.

We examined the effects of N additions on CH_4 consumption in soils from two tundra systems—moist tussock and wet meadow tundra. The soils were amended with amounts of N roughly equivalent to 50 and 100% of their net annual mineralization rates and incubated in the lab at 60% of water holding capacity and at either 5 or 15 °C (Kielland & Schimel, 1991). With these soils under these incubation conditions, the redox potentials stayed above 100 mV, which suggests that CH_4 production should be negligible. We found no significant effects on CH_4 consumption rates, suggesting that N availability does not affect CH_4 consumption in these soils. It remains to be determined whether this pattern holds true in other wetland and peatland soils. If so, it could suggest significant differences in the controls on CH_4 consuming populations from wetlands and uplands.

Oxygen Supply

Methane consumption in soils appears to be an obligatory aerobic process. In upland soils, where the CH_4 source is the atmosphere, O_2 is unlikely to be a major control on CH_4 consumption rates (Fig. 12–4). Any reduction in gas diffusion would limit both O_2 and CH_4 supply. Since atmospheric CH_4 concentrations are so low relative to the ability of organisms to use it, limiting gas diffusion should limit CH_4 supply to a greater degree than it limits O_2 supply.

In wetland soils, the case may be quite different (Fig. 12–5). While O_2 is supplied from the atmosphere, CH_4 is supplied from the subsurface anaerobic zone. In the surface zone, O_2 availability is controlled by its slow diffusion down and its rapid consumption in the soil. At the same time CH_4 diffusion from the anaerobic source zone may provide very high concentrations of CH_4. This could change the relative roles of CH_4 and O_2 limitation, and allow O_2 supply to act as an independent control on CH_4 consumption in the surface layer of wetland soils.

CONCLUSIONS

Over the last several years, great strides have been made in improving our understanding of the rates and controls on CH_4 dynamics in nature. We have also identified several major questions about these processes that we are only just starting to answer. We feel that there are four particularly pressing issues that soil scientists need to address:

1. Determination of quantitative relationships between C flow into the soil and CH_4 production rates.
2. Development of models of the partitioning of CH_4 transport between plants, diffusion, and ebullition.
3. Development of reliable estimates of the total soil CH_4 sink.
4. Determination and quantification of the actual mechanisms by which N inhibits CH_4 oxidation in the field.

ACKNOWLEDGMENTS

This work was supported by the USEPA, NSF, and NASA. We thank Knut Kielland and Knute Nadelhoffer for providing data and valuable discussion. We thank Lee Klinger for assisting with the collection of peat samples from the Hudson Bay Lowlands.

REFERENCES

Aselmann, I., and P.J. Crutzen. 1988. Global distribution of natural freshwater wetlands and rice paddies, their net primary productivity, seasonality and possible methane emissions. J. Atmos. Chem. 8:307–358.

Bédard, C., and R. Knowles. 1989. Physiology, biochemistry, and specific inhibitors of CH_4, NH_4^+, and CO oxidation by methanotrophs and nitrifiers. Microbiol. Rev. 53:68–84.

Born, M., H. Dörr, and I. Levin. 1990. Methane consumption in aerated soils of the temperate zone. Tellus 42(B):2–8.

Bouwman, A.F. (ed.). 1989. Soils and the greenhouse effect. John Wiley & Sons, New York.

Brown, A., S.P. Mathur, and D.J. Kushner. 1989. An ombitrophic bog as a methane reservoir. Global Biogeochem. Cycles 3:205–214.

Button, D.K. 1985. Kinetics of nutrient-limited transport and microbial growth. Microbiol. Rev. 49:270–297.

Cicerone, R.J., and R. Oremland. 1988. Biogeochemical aspects of atmospheric methane. Global Biogeochem. Cycles. 2:299–328.

Conrad, R. 1989. Control of methane production in terrestrial ecosystems. p. 39–58. In M.O. Andreae and D.S. Schimel (ed.) Exchange of trace gases between terrestrial ecosystems and the atmosphere. Wiley, New York.

Conrad, R. 1984. Capacity of aerobic microorganisms to utilize and grow on atnospheric trace gasses (H_2, CO, CH_4). p. 461–467. *In* M.J. Klug and C.A. Reddy (ed.) Current perspectives in microbial ecology. Am. Soc. Microbiol., Washington, DC.

Fung, I., J. John, J. Lerner, E. Matthews, M. Prather, L.P. Steele, and P.J. Fraser. 1991. Three dimensional model synthesis of the global methane cycle. J. Geophys. Res. 96:13 033–13 065.

Galchenko, V.F., A. Lein, and M. Ivanov. 1989. Biological sinks of methane. p. 39–58. *In* M.O. Andreae and D.S. Schimel (ed.) Exchange of trace gases between terrestrial ecosystems and the atmosphere. John Wiley and Sons, New York.

Harriss, R.C., D.I. Sebacher, and F.P. Day, Jr. 1982. Methane flux in the Great Dismal Swamp. Nature (London) 297:673–674.

Hutchinson, G.L., and G.P. Livingston. 1993. Use of chamber systems to measure trace gas fluxes. p. 63–78. *In* L.A. Harper et al. (ed.) Agricultural ecosystem effects on trace gases and global climate change. ASA Spec. Publ. no. 55. ASA, CSSA, and SSSA. Madison, WI.

Keller, M., T.J. Goreau, S.C. Wofsy, W.A. Kaplan, and M.B. McElroy. 1983. Production of nitrous oxide and consumption of methane by forest soils. Geophys. Res. Lett. 10:1156–1159.

Keller, M., W.A. Kaplan,a nd S.C. Wofsy. 1986. Emission of N_2O, CH_4, and CO_2 from tropical forest soils. J. Geophys. Res. 91:11791–11802.

Keller, M., M.E. Miter, and R.F. Stallard. 1990. Consumption of atmospheric methane in soils of central Panama: effects of agricultural development. Global Biogeochem. Cycles 4:21–28.

Khalil, M.A.K., and R.A. Rasmussen. 1990. Constraints on the global sources of methane and an analysis of recent budgets. Tellus 42(B):229–236.

Kielland, K., and J.P. Schimel. 1991. Temperature and moisture controls over carbon flux in arctic tundra soil. p. 21. *In* Global change and the biogeochemistry of radiative trace gases. Abstr. Int. Symp. Environ. Biogeochem, 10th, San Francisco, CA. 19–23 August.

Knowles, R. 1993. Methane: Processes of production and consumption. p. 145–156. *In* L.A. Harper et al. Agricultural ecosystem effects on trace gases and global climate change. ASA Spec. Publ. no. 55. ASA, CSSA, SSSA, Madison, WI.

Lindau, C.W., P.K. Bollich, R.D. Delaune, W.H. Patrick, Jr., and V.J. Law. 1991. Effect of urea fertilizer and environmental factors on CH_4 emissions from a Louisiana, USA rice field. Plant Soil 136:195–203.

Mariko, S., Y. Harazono, N. Owa, and I. Nouchi. 1991. Methane in flooded soil water and the emission through rice plants to the atmosphere. Environ. Exp. Bot. 31:343–350.

Mosier, A., D. Schimel, D. Valentine, K. Bronson, and W. Parton. 1991. Methane and nitrous oxide fluxes in native, fertilized and cultivated grasslands. Nature (London) 350:330–332.

Rosenfeld, S.A., and J.E. Brenchley. 1983. Regulation of glutamate and glutamine biosynthesis. p. 1–17. *In* K.M. Herrmann and R.L. Sumerville (ed.) Amino acids: Biosynthesis and genetic regulation. Addison-Wesley Publ., Co., Reading, PA.

Schmidt, E.L. 1982. Nitrification in soil. p. 253–288. *In* F.J. Stevenson (ed.) Nitrogen in agricultural soils. Agron. Monogr. 22. ASA, Madison, WI.

Schütz, H., P. Schröder, and H. Rennenberg. 1991. Role of plants in regulating methane flux to the atmosphere. p. 29–63. *In* T. Sharkey et al. (ed.) Trace gas emissions by plants. Academic Press, New York.

Seiler, W., R. Conrad, D. Scharffe. 1984. Field studies of CH_4 emissions from termite nests into the atmosphere and measurements of CH_4 uptake by tropical soils. J. Atmos. Chem. 1:171–187.

Steudler, P.A., R.D. Bowden, J.M. Melillo, and J.D. Aber. 1989. Influence of nitrogen fertilization on methane uptake in temperate forest soils. Nature (London) 341:314–316.

Valentine, D.W., E.A. Holland, and D.S. Schimel. 1992. pH dependence of CH_4 and CO_2 production in two contrasting Hudson Bay lowland wetlands. p. 45. *In* Global change and the biogeochemistry of radiative trace gases. Abstr. Int. Symp. Environ. Biogeochem., 10th, San Francisco, CA. 19–23 Aug. 1991.

Weast, R.C. 1976. Handbook of chemistry and physics. 57th ed. CRC Press, Cleveland, OH.

Whalen, S.C., W.S. Reeburgh, and K.A. Sandbeck. 1990. Rapid methane oxidation in a landfill cover soil. Appl. Environ. Microbiol. 56:3405–3411.

Whalen, S.C., W.S. Reeburgh, and K.S. Kizer. 1991. Methane consumption and emission by taiga. Global Biogeochem. Cycles 5:261–273.

Yarrington, M.R., and D.D. Wynn-Williams. 1985. Methanogenesis and the anaerobic microbiology of a wet moss community at Signy Island. p. 229–233. *In* W.R. Siegfried et al. (ed.) Antarctic nutrient cycles and food webs. Springer-Verlag, New York.

Yavitt, J.B., D.M. Downey, G.E. Land, and A.J. Sextone. 1990. Methane consumption in two temperate forest soils. Biogeochemistry 9:39–52.

13 Methane Emissions from Flooded Rice Amended with a Green Manure

Julie G. Lauren and John M. Duxbury

Department of Soils, Crops and Atmospheric Sciences
Cornell University
Ithaca, New York

Considerable interest in methane (CH_4) sources and sinks has arisen due to reports of a gradual increase in atmospheric CH_4 concentrations of 1% per year (Blake & Rowland, 1988). Because of the role of CH_4 in tropospheric and stratospheric chemistry and absorption of planetary infrared radiation, such an increase has implications for ozone (O_3) depletion and global warming. For example, Pearce (1989) estimates that atmospheric CH_4 levels account for 18% of current greenhouse warming, and will be the major greenhouse gas within 50 yr.

Flooded rice (*Oryza sativa*) has been identified as the most important source of anthropogenic CH_4, with estimates of annual emissions ranging between 50 to 170 Tg yr^{-1} and representing 21 to 25% of total emissions from all sources (Schütz et al., 1989; Sass et al., 1990). These levels are likely to increase as rice production expands to meet the needs of rising populations. Furthermore, management strategies utilized in flooded rice systems may enhance CH_4 emissions. For example, green manures, while contributing significantly to the sustainability of rice production in poorer countries where inorganic fertilizers are otherwise unavailable or expensive, may promote CH_4 production by providing C substrate for methanogenic bacteria. Schütz et al. (1989) and Yagi and Minami (1990) reported significant increases in CH_4 emissions with rice straw. However, no research has been conducted to determine CH_4 emissions from green manures.

The objectives of this experiment were to: (i) quantify and compare CH_4 emissions over a growing season from flooded rice with and without a green manure amendment; and (ii) determine the effect of soil type on CH_4 emissions from this system.

MATERIALS AND METHODS

A greenhouse experiment was established at the Guterman Bioclimatic Laboratory, Cornell University beginning in August 1991. Rice was grown to maturity (104 d) in flooded tubes filled with either of two soils, Lima (Glossoboric Hapludalf) or Kendaia (Aeric Hapluaquept). Each of the soils had the following treatments: soil only (S), soil plus rice (SR), soil plus green manure (SG), and soil plus both green manure and rice (SGR). All treatments were replicated three times.

Experimental units for the rice experiment were PVC tubes 10 cm in diameter by 25.4 cm long. Each tube was capped and sealed at one end with waterproof caulking. The appropriate soil was packed in each tube to a depth of 20.4 cm at a bulk density of 1.35 g cm^{-3} and flooded for approximately 5 wk prior to the start of the experiment.

Sesbania aculeata, a legume grown commonly as a green manure in South Asia, was cultured prior to the rice experiment. Pregerminated seeds were sown in potting mix and watered with Peters nutrient solution (W.R. Grace, New Jersey) periodically. After 62 d, the *Sesbania* tops were harvested and thoroughly mixed using a feed chopper. Subsamples for determination of moisture, N and C contents were removed from the central batch. A Europa Scientific Roboprep C–N analyzer (Europa Scientific Ltd., Crewe, United Kingdom) was utilized to determine total N and C contents of the added *Sesbania* as well as the Lima and Kendaia soils. Results from the analysis are presented in Table 13–1.

Six days before rice transplanting, fresh *Sesbania* at 25 g tube^{-1} (equivalent to 30 mt fresh weight ha^{-1}) was incorporated into the top one-third of each of the green manure treatment tubes. Transplantation of the rice was accomplished by placing two pregerminated seeds of Ai-nan-tsao rice on the soil surface of each of the planted tubes. Tubes were maintained in a flooded condition for the duration of the experiment.

The closed chamber method was used to measure CH_4 fluxes from the tubes over the growing season. Collection chambers, each with a stoppered vent for pressure equilibration and a half-hole septum sampling port were attached to the tops of the tubes with rubber bands cut from tire inner tubes (Fig. 13–1). Chamber volumes were 1 or 2 L, depending on the height of the rice plants. During sampling, a 10-mL gas sample was withdrawn from each chamber initially and subsequently at 10-min intervals up to 20 min with disposable glass syringes. Syringe needles were inserted into rubber stoppers

Table 13–1. Carbon and N contents for *Sesbania* green manure and soils used in the experiment.

	Total N	Total C
	%	
Sesbania	1.6	49.5
Lima soil	0.16	1.9
Kendaia soil	0.23	2.3

to prevent leakage prior to analysis. The chambers were removed after each collection period. Internal chamber temperatures after gas sampling were not more than 2 °C higher than outside the chambers.

Gas samples were analyzed immediately after collection using an Aerograph Model 600 HyFi gas chromatograph with a flame ionization detector (Wilkens Instrument & Research, Inc., Walnut Creek, CA). A 5-mL sample was injected into a 0.6-mL sampling loop and separated on a Chromosorb 102 column (Applied Sciences Laboratories, Inc., State College, PA) (0.03-m i.d. by 3.7 m) with 30 mL min^{-1} of N_2 carrier gas at ambient temperature. The gas chromatograph was calibrated daily with standards ranging from 4 to 4000 ppmv.

Flux measurements were made roughly every 2 d on the three replicate tubes for each treatment. On sampling days, flux measurements were taken early in the morning and once again in the afternoon. These times correspond to minimum and maximum flux events (See Results and Discussion). Daily maximum, minimum and current air temperatures were also recorded on sampling days.

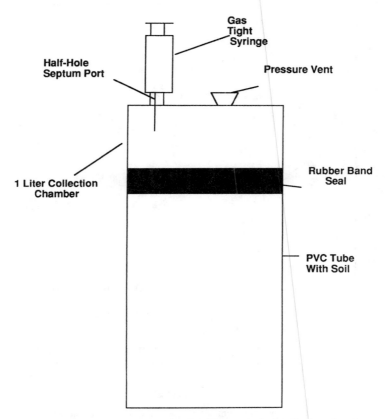

Fig. 13-1. Schematic of an experimental unit with an attached collection chamber.

RESULTS AND DISCUSSION

Several investigators have demonstrated a strong correlation between temperature and CH_4 emissions (Seiler et al., 1984; Holzapfel-Pschorn & Seiler, 1986; Yagi & Minami, 1990). Given the observed differences in daily maximum and minimum temperatures as displayed in Fig. 13–2, we anticipated different fluxes depending on the temperature and the time of sampling. Thus in characterizing CH_4 emissions from these treatments, we report a range of minimum and maximum fluxes corresponding with the minimum and maximum temperatures of each sampling day.

Many of the CH_4 emission patterns recorded during early afternoon samplings (maximum temperatures) showed nonlinear increases indicating that ebullition processes dominated emissions. Also bubbles were observed on the surface of both the soil and in the floodwater during these samplings, especially for the SG treatments. In contrast, few bubbles were seen during minimum temperature samplings, and CH_4 concentrations generally in-

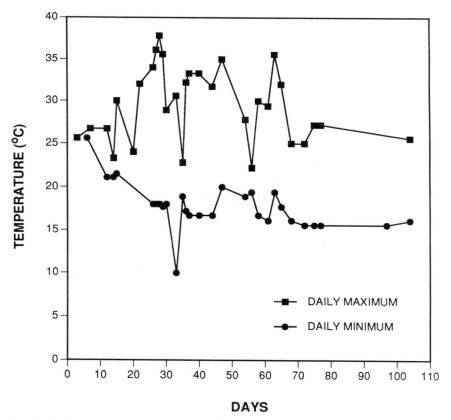

Fig. 13–2. Daily maximum and minimum air temperatures in the greenhouse during the experimental period.

creased linearly during the collection period, suggesting that diffusive or plant-mediated transport processes dominated emissions at these times.

Short-term measurements of CH_4 concentrations following bubble release violate the linearity assumption for a flux calculation. As Fig. 13–3 demonstrates, measurements collected over a longer period showed that the overall trend was linear. A similar observation was made by Seiler et al. (1984). Rather than discard data sets dominated by bubble release, a linear model was assumed to calculate fluxes when obvious ebullition had occurred. We believe that it was important to include bubble release in the flux measurements, since exclusion would have ignored a significant transport process and seriously underestimated emissions.

Green Manure Effects

Figure 13–4 displays the results from the planted treatments with and without *Sesbania* green manure. Consistently higher CH_4 fluxes were measured in the SGR treatments relative to the SR treatments, for both minimum and maximum temperature samplings and both soils. Highest CH_4 fluxes were observed during the first 30 to 35 d, which correspond to the most active period of *Sesbania* decomposition. Emission differences were maintained through 60 d after incorporation (DAI) after which few differences were noted, presumably because readily available C from the *Sesbania* was ex-

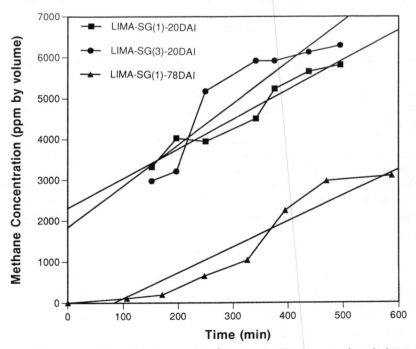

Fig. 13-3. Demonstration of the long-term linear increase in CH_4 concentrations during emissions dominated by ebullition processes.

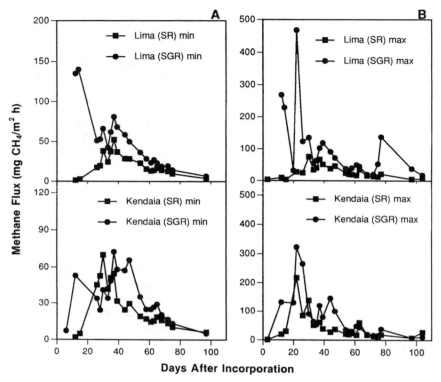

Fig. 13–4. Time–course CH_4 flux values for the Lima and Kendaia soils between planted treatments with and without *Sesbania*, (A) minimum and (B) maximum temperature samplings.

hausted. Thus these data confirm that a green manure addition will enhance CH_4 emissions in flooded rice by providing C substrate to methanogenic bacteria.

A second peak in CH_4 production at 70 to 80 DAI is evident in the Lima (maximum) plot for the SGR treatment. As observed by Schütz et al. (1989) and Lindau et al. (1991), the increase in emissions at this time may be attributed to C supplied by root lysis or exudation from rice after flowering. Similar patterns expected in the other planted treatments were not observed, which may have been due to less frequent sampling at this critical time.

Soil Effects

Figure 13–5 compares CH_4 emissions between the two soils for each treatment. Fluxes from the S treatment were negligible throughout the experiment and are not reported. Few differences in CH_4 flux were observed between the Lima and Kendaia soils after 30 DAI, while earlier in the experiment, higher CH_4 emissions were noted for the Kendaia soil in the SR (minimum and maximum) and SG (minimum) treatments. These results are consistent with higher total soil C levels in the Kendaia soil relative to the

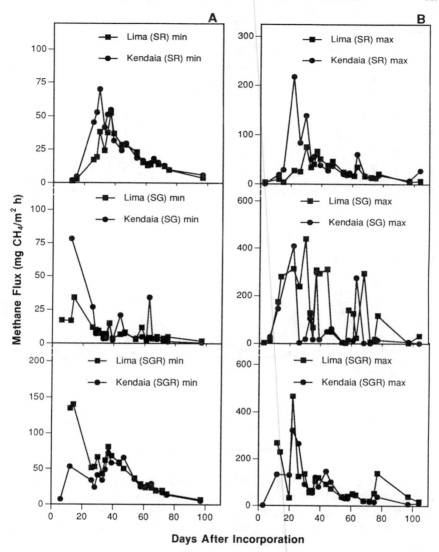

Fig. 13-5. Comparisons between the Lima and Kendaia soils of CH_4 fluxes from the SR, SG, and SGR treatments; (A) minimum and (B) maximum temperature samplings.

Lima soil (see Table 13-1). Few flux differences were apparent in the SG (maximum) and SGR (maximum) treatments throughout the experiment. High added C and ebullition factors may have combined to swamp out any soil differences.

Greater CH_4 emissions were evident in the first 30 d for the Lima soil compared to the Kendaia in the SGR (minimum) treatment. This result is unusual given the higher soil C levels in the Kendaia soil. Perhaps an interaction with the green manure addition and rice would explain this result.

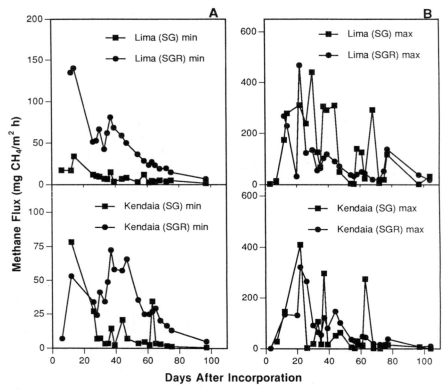

Fig. 13–6. Time–course CH$_4$ flux values for the Lima and Kendaia soils between the planted and unplanted *Sesbania* treatments, (A) minimum and (B) maximum temperature samplings.

Rice Plant Effects

Research by Seiler et al. (1984), Holzapfel-Pschorn and Seiler (1986), and Schütz et al. (1989) indicates that much of the observed CH$_4$ emissions from flooded rice are transported via diffusion through the rice plant. Data obtained from the minimum temperature samplings in our experiment are in agreement with this statement (Fig. 13–6A), with CH$_4$ emissions from the planted SGR treatments about two to four times the unplanted SG treatments. Likewise, total hourly CH$_4$ emissions from the SR treatments ranged from 1.3 to 3.7 g CH$_4$ m^{-2} at the end of the experiment, while no significant fluxes were detected from the unplanted S treatments.

However, a different picture emerges from the results of the maximum temperature samplings between the SG and SGR treatments (Fig. 13–6B). Emissions from the unplanted SG treatments were similar and in some cases larger than the planted SGR treatment emissions. We attribute the high fluxes from the SG treatments to ebullition processes and high C loading from the green manure addition. These results indicate that ebullition can be just as effective a CH$_4$ transport mechanism as plant-mediated diffusion, especial-

Table 13-2. The range of CH_4 emissions integrated over the growing season for the SR, SG, and SGR treatments. Values for the S treatments were negligible.

Treatment	Minimum		Maximum		Mean	
	Lima	Kendaia	Lima	Kendaia	Lima	Kendaia
			—————— g CH_4 m^{-2} ——————			
SR	34	44	55	98	45	71
SG	20	27	296	178	158	103
SGR	93	69	204	176	148	122

ly under conditions of warm temperatures and high C levels. Also, this suggests that to adequately measure CH_4 flux rates from these systems, more research is required to integrate bubble release into flux measurements. Past research efforts have focused on determining the concentration of CH_4 in bubbles (Holzapfel-Pschorn & Seiler, 1986) or quantifying the reserve of CH_4 trapped in the soil as bubbles (Seiler et al., 1984), but little work has been done to measure bubble-dominated fluxes. While we have chosen to include bubble release in our flux measurements assuming the linear model, it should not be considered the only approach.

Total Methane Emissions

Total quantities of CH_4 emitted during minimum and maximum temperature sampling times by each treatment are presented in Table 13-2. The average of these emission ranges are presented in the third column.

Methane emissions from the green manure treatments were substantially greater than those reported by either Schütz et al. (1989) (33–38 g CH_4 m^{-2}) or Yagi and Minami (1990) (27–45 g CH_4 m^{-2}) for comparable application rates of rice straw. Disparities in CH_4 emissions between a green manure and rice straw may reflect differing decomposition rates and by-products, measurement differences (e.g., including vs. excluding bubble release in flux measurements) or differing conditions of soil anaerobiosis. More research is necessary to determine the true cause.

Based on the mean CH_4 emission rates from the green manure treatments, it was determined that 24 to 37% of the added *Sesbania* C was emitted as CH_4. This result suggests that green manure C additions to flooded rice could have a significant impact on atmospheric CH_4 levels. More data are necessary to fully establish the quantities of green manure or plant residue C emitted as CH_4 from flooded rice systems.

CONCLUSIONS

1. Application of *Sesbania* green manure to flooded rice increased CH_4 emissions by two to three times, relative to rice without green manure.

2. Higher soil C levels in the Kendaia soil relative to the Lima soil resulted in elevated CH_4 fluxes initially for the minimum temperature samplings. Ebullition and added C appeared to swamp out any soil effects in the green manure treatments at the maximum temperature sampling.

3. Transport of CH_4 was predominantly diffusive through the rice plant at minimum temperatures. At maximum temperatures, however, ebullition processes are just as important for CH_4 transport as plant-mediated diffusion, especially when added C levels were high. Future field efforts to characterize CH_4 fluxes from flooded rice amended with green manures or plant residues should consider the importance of ebullition under these conditions.

SUMMARY

Green manures contribute significantly to the sustainability of flooded rice production in poorer countries. An unfavorable side effect of green manure use may be the promotion of CH_4 emissions and their effects on global warming. A greenhouse experiment was performed to quantify the relative proportions of CH_4 coming from soil, rice plants or a green manure (*S. aculeata*). Results indicated that application of *Sesbania* green manure to flooded rice increased CH_4 emissions by two to three times, relative to rice without green manure. Differing CH_4 fluxes were noted between unamended soils, but ebullition and added C overwhelmed any soil effects in the green manure treatments. The data also confirmed that rice plants are an important conduit for CH_4 release; however, under conditions of high temperature and added C, ebullition processes were just as important for CH_4 transport as plant-mediated diffusion.

REFERENCES

Blake, D.R., and F.S. Rowland. 1988. Continuing worldwide increase in tropospheric methane, 1978 to 1987. Science (Washington, DC) 239:1129–1131.

Holzapfel-Pschorn, A., and W. Seiler. 1986. Methane emission during a cultivation period from an Italian rice paddy. J. Geophys. Res. 91:11 803–11 814.

Lindau, C.W., P.K. Bollich, R.D. Delaune, W.H. Patrick, and V.J. Law. 1991. Effect of urea fertilizer and environmental factors on CH_4 emissions from a Louisiana, USA rice field. Plant Soil 136:195–203.

Pearce, F. 1989. Methane: The hidden greenhouse gas. New Scientist 6 May, p. 19–23.

Sass, R.L., F.M. Fisher, and P.A. Harcombe. 1990. Methane production and emission in a Texas rice field. Global Biogeochem. Cycles 4:47–68.

Schütz, H., W. Seiler, and R. Conrad. 1989. Processes involved in formation and emission of methane in rice paddies. Biogeochemistry 7:33–53.

Seiler, W., A. Holzapfel-Pschorn, R. Conrad, and D. Schaiffe. 1984. Methane emission from rice paddies. J. Atmos. Chem. 1:241–268.

Yagi, K., and K. Minami. 1990. Effect of organic matter application on methane emission from some Japanese paddy fields. Soil Sci. Plant Nutr. 36:599–610.

14 Rapid, Isotopic Analysis of Selected Soil Gases at Atmospheric Concentrations

Paul D. Brooks and Donald J. Herman

Department of Soil Science
University of California
Berkeley, California

Gary J. Atkins, Simon J. Prosser, and Andrew Barrie

Europa Scientific Ltd.
Cheshire, United Kingdom

The isotopic abundance measurement of ^{15}N in N_2O, N_2, and ^{13}C in CH_4 and CO_2 is useful not only in the measurement of small fluxes of these trace gases but also in the determination of their sources and metabolic pathways by which they are formed. Isotopic analysis of atmospheric levels of N_2O, CH_4, and CO_2 by conventional mass spectrometry requires liters of sample, large glass manifolds, and hours of preparation per sample (Yoshida et al., 1983; Lowe et al., 1991; Mook et al., 1983). Mulvaney and Kurtz (1982) published a method for analyzing $^{15}N_2O$ and $^{15}N_2$ from soil gas, but this still required a tedious preparation and elaborate manifold. Strong et al. (1987) described a method to convert N_2 and N_2O to NO_3, but this required considerable wet chemistry.

Fluxes of N_2O, CH_4, and CO_2 from soil are often measured by placing a vented cover over soil, and taking samples for up to 1 h (Hutchinson & Mosier, 1981). The flux of gas can be calculated by a linear increase in its concentration. Only small (20–60 mL) aliquots of sample are taken to avoid too much air entering the cover vent. Because of the spatial variability, large numbers of samples must be taken. Usually only small samples (0.1–20 mL) are available for analysis in metabolic studies. Therefore a method of analyzing the isotopes of N_2O and CH_4 from 13-mL samples at atmospheric concentrations was developed, with precision suitable for tracer work.

Initial work with a packed column interfaced to an isotope ratio mass spectrometer (IRMS; Tracermass, Europa Scientific Ltd., Europa House, Electra Way, Crewe, Cheshire, CW1 1ZA United Kingdom) showed that there was insufficient signal for precise measurement of N_2O at atmospheric concentration. Gutherie and Duxbury (1978) concentrated N_2O onto a molecular sieve, then baked it off into a gas chromatograph. Here we describe a method where N_2O is sorbed on a zeolite in a low dead volume trap assembly. It is then thermally desorbed into a slow carrier flow and passed to the IRMS with increased sensitivity as a result. When N_2O is analyzed the $^{15}N_2$ can be measured on the same sample. Data using this method was published by Brooks and Atkins in 1992. The method has subsequently been automated and developed for CH_4 analysis as well.

A range of sample vials suitable for sample collection and storage were also evaluated. The sample vials were tested for N_2O concentration as shipped, and leakage of N_2O during long-term storage.

MATERIALS AND METHODS

Manual Method

Twenty milliliters of sample is injected through a 13-mL loop on Valve 1 and allowed to equilibrate to atmospheric pressure so pressure transients from the injection are minimized and the injection volume is consistent (Fig. 14-1). The barometric pressure is continuously monitored in the hallway adjacent to the mass spectrometer lab, and a large change in 1 h is 760.7 to 758.2 mm Hg, or a 0.3% change. Since the instrument is recalibrated every hour, this change is considered negligible. The valve is then turned, and the 13 mL injected into a stream of He flowing at 60 mL min^{-1}, passed through $Mg(ClO_4)_2$ and Carbosorb (BDH Chemicals Ltd., Poole, United Kingdom) to remove H_2O and CO_2, and then through Valve 2 to a 50- by 0.65-mm column at ambient temperature packed with 100–120 mesh molecular sieve 13X. This column traps the N_2O, within 1 min the N_2, O_2, Ar, and trace gases NO, CH_4, and CO pass to a 1-m by 3.1-mm molecular sieve 5A column at 0 °C. Valve 2 is then turned so that the N_2O remains trapped.

The N_2 is separated from O_2 on the 1-m column, and passes through Valve 3 to the split valve, where about 0.1 mL min^{-1} of the carrier stream goes into the mass spectrometer. The mass spectrometer is tuned to measure masses 28, 29, and 30 as the N_2 elutes from the column. Oxygen elutes before N_2 and CO after. Small amounts of NO might cause the mass 30 peak to tail slightly, but this has not been observed.

Valve 3 is turned, the 50-mm column put in an oven at 135 °C, and the mass spectrometer automatically retuned to masses 44, 45, and 46. A low flow rate of 0.3 mL min^{-1} of He carries the N_2O released by the higher temperature into the split inlet of the mass spectrometer.

After this analysis the valves are reset and the 50-mm column cooled before the next analysis cycle. A sample can be run every 12 min. Replicate analyses of air were made to determine the instrument's precision.

To test the instrument a 25-cm diam. PVC ring was hammered into soil at the University of California-Berkeley and 1 L of water added to the dry

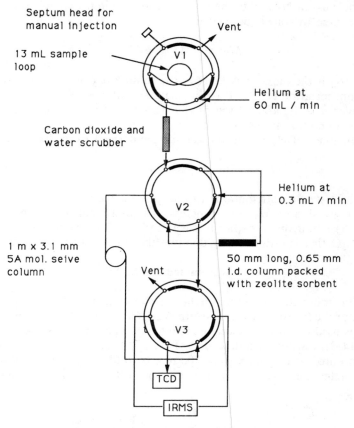

Fig. 14–1. Diagram of manual gas GC inlet to mass spectrometer.

soil. One hour later 200 mL of 29 μg mL^{-1} of NO_3–N at 50% ^{15}N enrichment was added to the soil (equivalent to 1.0 kg ha^{-1}), and a 8.3-L cover placed on the ring. Twenty-milliliter gas samples were removed at 15, 27, 39 and 51 min, and analyzed for $^{15}N_2$ and $^{15}N_2O$.

The fraction of N_2O in a gas sample attributable to the spike may be derived from the following

$$xE_S + yE_N = E_M \qquad [1]$$

that rearranges to

$$x = (E_M - E_N)/(E_S - E_N) \qquad [2]$$

where E_S is the enrichment (in atom %) of the spike, E_N is the enrichment of the native N source, E_M is the enrichment measured on the mass spectrometer, x is the fraction of N_2O contributed by the spike, and y is the fraction of N_2O contributed by the native N source.

The mass of N_2O-N in the gas above the soil in the chamber, expressed per unit area of soil, is derived from

$$M_M = C_M VN\,ART^{-1} \qquad [3]$$

where M_M is the mass of N_2O-N in the sample, C_M is the measured total N_2O concentration (V^{-1}), V is the chamber volume, N is 28 g of N_2O-N per mole, A is the area of the ring, R is 8.3143 J mol^{-1} degree^{-1}, and T is the temperature in Kelvin.

The mass of N_2O-N in the gas sample derived from the spike (M_S) is

$$M_S = xM_M \qquad [4]$$

The N_2O flux attributable to the spike is computed by performing a linear regression of M_S upon time over the linear portion of the response curve. Assuming that the spike is in equilibrium with the native soil N source, total N_2O flux is estimated by dividing the spike N_2O flux by x.

Automated Method

The complete automatic method is shown in Fig. 14–2. A concentric needle probe (Europa House, Electra Way, Crewe, CW1 1ZA, United Kingdom) is used to flush the sample from a container into a He carrier stream (Fig. 14–3). Carbon dioxide and H_2O are chemically removed, and the N_2O sorbed onto a zeolite 13X trap (200 mm by 0.65 mm, 100 to 120 mesh) at ambient temperature. Dinitrogen, O_2, and other sample components are

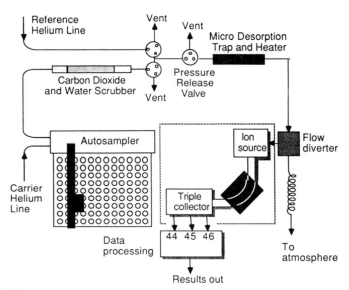

Fig. 14–2. Automatic sampler, GC system, and isotope mass spectrometer for automated N_2 and N_2O analysis.

passed to the IRMS where the N_2 is analyzed for ^{15}N. The carrier flow is reduced to approximately 0.5 mL min $^{-1}$ and the N_2O is analyzed for ^{15}N. The carrier flow is reduced to approximately 0.5 mL min $^{-1}$ and the N_2O thermally desorbed as in the manual method and measured at masses 44, 45, and 46. (Fig. 14–2).

For analysis of CH_4, a tube furnace containing platinized Cu (II) oxide and reduced Cu granules was added (Fig. 14–4). The CO_2 resulting from CH_4 combustion was sorbed onto the zeolite, prior to thermal desorbtion and isotopic analysis. Standards of 385, 308, 231, 154, and 77 $\mu L\ L^{-1}$ were prepared by successive dilution of a 5% mixture (Phase Separations, Queensferry, Wales, United Kingdom). Five measurements of each standard were run against themselves as isotopic references and set to a nominal 0.00‰ ^{13}C. Blank values for Exetainers containing only N_2 were automatically subtracted from each sample analyzed. For atmospheric samples, a gas chromatograph (GC) clean-up stage will be required to purify CH_4 prior to combustion, this is indicated in Fig. 14–4 but was not used for results reported here.

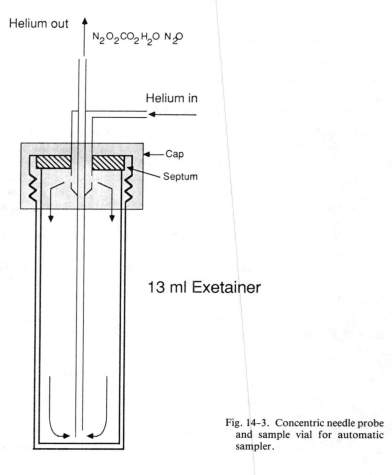

Fig. 14–3. Concentric needle probe and sample vial for automatic sampler.

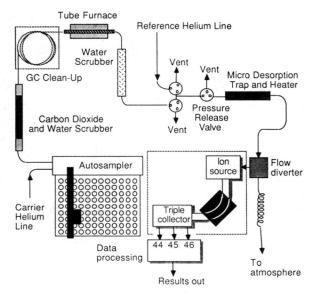

Fig. 14–4. Automatic sampler, GC system, furnace and mass spectrometer for CH_4 analysis.

Evaluation of the Containers for Sample Storage

The following containers were tested:

1. 16- by 125-mm, 17-mL screw-cap culture tubes with Hungate septa (no. 2047-11 600) and caps (no. 2047-16 000, Bellco Glass Inc., P.O. Box B, 340 Edrudo Road, Vineland, NJ 08360, USA).
2. 16- by 100-mm, 14-mL screw-cap culture tubes with Hungate septa (no. 2047-11 600) and caps (no. 2047-16 000, Bellco Glass Inc.).
3. Exetainers, 13 mL, (Europa Scientific, Europa House, Crewe, CW1 1ZA, United Kingdom).
4. Vacutainers, 20 mL, (no. 6433, Becton-Dickinson, Rutherford, NJ 07070, USA).
5. Crimp-top 10-mL serum vials (no. 223 738) with red rubber septa (no. 224 124, Wheaton, 1501 North Tenth Street, Millville, NJ 08332, USA).

The following treatments were applied to two replicates of the Exetainers and Vacutainers and five replicates of each of the remaining container types:

1. To measure initial N_2O degassing, the containers were placed under 90 kPa of vacuum for 1 min, pressurized with N_2 for 1 min, evacuated for another minute, and finally pressurized with N_2 to 7 kPa for 1 min.
2. The first preparation to test integrity during storage was to vent the vial with an open syringe needle, flush with 55 mL of a 0.5 μL L^{-1}

Table 14–1. Replicate analysis of air for ^{15}N, N_2, and N_2O.

	Atom % ^{15}N	
Replicate	N_2O	N_2
1	0.36064	0.36628
2	0.36442	0.36631
3	0.36218	0.36628
4	0.36767	0.36629
5	0.36675	0.36630
6	0.36412	0.36631
7	0.36543	0.3629
mean	0.3645	0.36629
SD	0.0025	0.00001

N_2O standard, remove the vent needle, and finally pressurize with an additional 5 mL of the standard.

3. The second preparation to test integrity during storage was to evacuate as in Treatment 1. Finally, the vial was filled to volume with a 1 μL L^{-1} N_2O in N_2 standard, then pressurized with an additional 5 mL of standard.

After vials were prepared as above, they were stored on a laboratory bench top for 14 d. Three-milliliter aliquots were withdrawn from each vial, injected into a 0.5-mL sample loop, and assayed for [N_2O] by an electron capture detector.

Dilution of the added gas sample may result from residual air in the vial or degassing or from leakage across the septum, usually through a syringe needle hole. The dilution of the gas sample was computed by the following relationship

$$xA + yS = M \qquad [5]$$

that rearranges to

$$x = (M - S)/(A - S) \qquad [6]$$

in which x is the fraction of measured N_2O contributed by ambient air, y is the fraction of measured N_2O contributed by the added standard, M is the measured [N_2O], S is the [N_2O] injected into the vial, and A is ambient [N_2O], measured in our laboratory to be 0.305 μl L^{-1}.

RESULTS AND DISCUSSION

Manual Method

Replicate analysis of air is shown in Table 14–1. Precision for N_2 was similar to that of many dual inlet mass spectrometers, precision for N_2O sufficient for tracer studies. The result of N_2O flux experiment is shown in

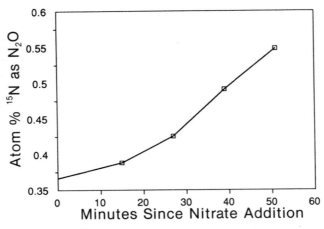

Fig. 14–5. Increase in the % ^{15}N of N_2O in a cover over soil treated with 0.5 kg ha^{-1} NO_3-N at 50% ^{15}N.

Fig. 14–5. There was no measurable increase in the concentration of N_2O, however, there was a readily measurable increase in its isotope ratio. This was calculated from Eq. [2], [3], and [4] as 400 μg N_2O-N ha^{-1} h^{-1}. Rates as low as 40 μg ha^{-1} h^{-1} could be measured by this method.

Automatic Method

Isotopic abundance measurements for ^{15}N in N_2O and ^{13}C in CH_4 are shown in Tables 14–2 and 14–3 respectively. Variation in the observed results for CH_4 are due to two factors, (i) large blanks caused by trace hydrocarbons in the Exetainer that combust with the CH_4 to form CO_2, and (ii) a constant bleed at masses 44, 45, and 46 from the furnace packing, both of

Table 14–2. Summary of N_2O results using the automated method.

Concentration	Atom % ^{15}N	SD
μL L^{-1}		
77	0.36609	0.0006
39	0.36574	0.0006
15	0.36610	0.0016
Atmosphere	0.34035	0.021

Table 14–3. Summary of CH_4 results using the automated method.

Concentration	δ^{13}C	SD ($n = 5$)
μL L^{-1}	per mil	
385	−0.376	0.42
308	−0.306	0.50
231	−0.296	0.47
154	0.836	0.47
77	0.386	0.85

Table 14–4. Mean concentrations or dilution rates measured in three treatments applied to five container types.

Container	Treatment		
	Initial degassing [N$_2$O]	Dilution after Preparation 2	Dilution after Preparation 3
	μL L^{-1}	——————— % ———————	
Screw-cap, 16 by 125 mm	0.016a*	11a	7.4a
Screw-cap, 16 by 100 mm	0.009a	3.8a	8.3a
Exetainers	0.126b	9.1a	6.9a
Vacutainers	2.102c	51b	17a
Serum vials	0.237d	56b	63b

* Results tagged with the same letter within a treatment are not significantly different at the 0.05 probability level.

which contribute to CO_2 that desorbs along with that originating from the CH_4 sample. Separation of CH_4 from other hydrocarbons using a GC clean-up column prior to the furnace should minimize the hydrocarbon blank, but has not been attempted yet. Although the precision with the automated system isn't as good as with the manual system, the automated system is useful in experiments where large numbers of samples need to be processed. Further work is needed to improve the precision of the automated method.

Evaluation of the Containers for Sample Storage

A one-way ANOVA with Tukey's multiple range comparison was performed on the results of each treatment to test for differences among container types with respect to initial degassing of N_2O, and dilution from extraneous sources of the stored sample following two methods of container preparation. Results are shown in Table 14–4.

1. Leakage through and degassing of N_2O from the red rubber serum stoppers used to seal Vacutainers and serum vials made them unsuitable as sample containers. Exetainers initially degas significant levels of N_2O; however, after degassing adequate flushing they may safely be used. Between degassing and leakage, serum vials are unsuitable. Screw-cap culture tubes with Hungate septa of all the containers tested most effectively exclude extraneous N_2O.

2. When sample containers are flushed with excess sample, then topped off with sample, Exetainers and screw-cap tubes allow minimal dilution. The high dilution observed with Vacutainers may have resulted from poor mixing dynamics during flushing.

3. When containers have been evacuated prior to sample injection, low dilution rates were observed among all container types except serum vials.

CONCLUSION

Utilizing equipment described herein with tracer ^{15}N and ^{13}C it is now possible to measure the production or fluxes of the trace gases CH_4 and N_2O in field experiments. Processing large numbers of samples by the automated system and method described should improve estimates made of fluxes of trace gases and thus enhance our understanding of their production.

Screw-cap tubes and Exetainers exhibited the most desirable characteristics in maintaining sample integrity. Screw-cap tubes and Exetainers may be prepared simply by flushing with an excess of sample, since there is no apparent advantage in first evacuating them. When collecting samples standards should be injected into separate sample or storage containers and assayed with the unknowns so that the results may be corrected for leakage.

ACKNOWLEDGMENTS

This research was supported in part by NSF grant DIR-8907131.

REFERENCES

Brooks, P.D., and G.J. Atkins. 1992. Rapid isotopic analysis of trace gases at atmospheric levels. Isotopenpraxis. (in press.)

Gutherie, T.F., and J.M. Duxbury. 1978. Nitrogen mineralization and denitrification in organic soils. Soil Sci. Soc. Am. J. 42:908–912.

Hutchinson, G.L., and A.R. Mosier. 1981. Improved soil cover method for field measurement of nitrous oxide fluxes. Soil Sci. Soc. Am. J. 45:311–316.

Lowe, D.C., C.A.M. Brenninkmeijer, S.C. Tyler, E.J. Dlugkencky. 1991. Determination of the isotopic composition of atmospheric methane and its application in the antarctic. J. Geophys. Res. 96(D8):15 455–15 467.

Mook, W.G., M. Koopmans, A.F. Carter, and C.D. Keeling. 1983. Seasonal, latitudinal, and secular variations in the abundance and isotopic ratios of atmospheric carbon dioxide. 1. Results from land stations. J. Geophys. Res. 88(C15):10 915–10 933.

Mulvaney, R.L., and Lt.K. Kurtz. 1982. A new method for determination of ^{15}N-labeled nitrous oxide. Soil Sci. Soc. Am. J. 46:1178–1184.

Strong, W.M., E.R. Austin, L.S. Holt, and R.J. Buresh. 1987. Determination of the combined nitrogen-15 content of dinitrogen and nitrous oxide in air. Soil Sci. Soc. Am. J. 51:1344–1350.

Yoshida, N., and S. Matsuo. 1983. Nitrogen isotope ratio of atmospheric N_2O as a key to the global cycle of N_2O. Geochem. J. 17:231–239.

SUBJECT INDEX

Abiotic N oxide production, 80, 102
Acetate, 171
  substrate, 159
Acetotrophs, 146
Acetylene, 152
  block, 133
  inhibition, 82
Acidic deposition, 79, 97
Activation energy, 147, 151
Aerenchyma, 173
Aerosols, 13–14
Agriculture
  contribution to trace gases, 7–12, 45, 95
  contribution to climate change, 7–12
  future scenarios, 14–15
  differential effects on trace gases, 12
Air density fluctuations, 54
Aircraft-based, 46, 47, 56, 57, 104
Alaska, 179
Alcaligenes, 81, 85
Allelopathic compounds, 83
Allylthiourea, 152, 153
Ammonia monooxygenase, 152
Ammonia (NH_3), 3, 12, 15, 19, 26, 28, 30, 63
Ammonium, 178, 179
  fertilization, 99, 133, 141
AMO, 152
Anaerobic bacteria, 159
Anthropogenic sources, 95–106, 126–128, 130, 158
Aquifer(s), 121
Archaeobacteria, 145
Arctic, 173
Aspergillus flavus, 81
Atmospheric aerosols, 13–14
Atmospheric chemistry, 95–97, 167
Atmospheric oxidants, 79, 97
Atmospheric stability, 23, 28, 29, 96
Automated soil gas isotope ratio measurement, 196

Background concentration, 21, 26, 28, 41, 95
BES, 147, 152, 153
Biogeochemistry, 19
Biological Nitrogen fixation (BNF), 9
Biomass burning, 97–98, 102–103, 158
Boundary layer, 23, 25, 28, 30, 34–41, 59
Bromoethane sulfonate, 147, 152, 153

Candida, 151
C/N ratio, 116, 179
Carbon
  availability, 86–88, 171
  storage, 8, 15
  substrate, 183, 188

Carbon dioxide (CO_2), 1–9, 19, 21, 22, 24, 26, 27, 31–38, 40, 157
  assimilation, 22, 37
  concentration, 1–15, 31–38
  drawdown, 34, 37
  fertilization, 8
  isotope measurement, 197
Carbon disulfide (CS_2), 12, 14
Carbonyle sulfide (COS), 12, 14
CBL budget methods, 34–41
Chamber methods
  biological disturbance, 70, 21
  closed chamber, 20, 21, 65–76, 134, 184
  concentration effects, 21, 68–69, 71–74
  comparison of types, 65, 76
  flux models
    linear vs. nonlinear, 68–69, 71–74
    significance testing, 71–73, 76
    verifying model fit, 71–73, 76
  improving sensitivity, 73–74
  minimum detectable flux, 73
  pressure disturbance, 67–68
  site disturbance, 69–71
  systems/techniques, 20–22
  temperature disturbance, 66–67
Chemoautotrophic nitrification
  biochemical pathway, 82
  cellular-level control, 83
  gaseous products, 82–83
  importance to N cycle, 81
  $NO:N_2O$ emissions ratio, 82–83
  species diversity, 81, 124, 127
Chemodenitrification, 80, 115, 117
Chlorate inhibition, 82
Chlorofluorocarbons (CFC), 1–5, 95
Chlorofluoromethanes, 157
Chromatography
  column, 194, 196, 197
Circular plot, 29
Closed-path analysers, 27
Clouds, 13–14
Coefficient of variation, 72
Competition, 180
  for soil NH_4^+, 81, 117
  substrates, 146
Conditional sampling, 27
Consortium, 152
Consumption/deposition of N oxides, 63, 71, 79, 85, 88, 102–103
Convective boundary layer (CBL), 26, 34–41
Convective mixing, 34
Corn, irrigated, 70, 135, 138–142
Correlation coefficients, 48
Cospectral characteristics, 50
Covariance, 56

Crop residue, 89
Cultivation, 142

Damping of gas fluctuations, 27
Deforestation, 8, 14
Denitrification, 10–12, 121–132, 134, 138
  biochemical pathway, 86
  cellular-level control, 85–88
  controls, 123–125
  gaseous products, 86, 88
  importance to N cycle, 85
  $N_2O:N_2$ emissions ratio, 86, 88, 125, 129
  $NO:N_2O$ emissions ratio, 86, 88
  species diversity, 85–86
Detection limit, 24, 47
Detrended fluxes, 51
Diffusion, 68–69, 175, 178, 180
Dimethyl ether, 152, 153
Dissipation technique, 48
Diurnal cycle, 112–115

Ebullition, 174, 186–187, 190, 192
Eddy accumulation, 27, 28, 47
Eddy correlation techniques, 26, 27, 46, 47
Eddy diffusivity, 23, 24
Eddy flux, 26, 27
Energy balance, 34, 56
Entrainment, 34, 37
Ethanol, 171, 173
Evaporation, 26, 31–34

Fen, 171
Fermentation, 171
Fertilization, 179, 180
Fertilizer, 161, 162
Fetch, 23, 28, 29, 55
Flux
  chambers, 63–76, 111
  diffusive, 29, 68, 69
  divergence, 47
  gas, 21, 28, 30, 33, 34
  local, 37–40, 63, 74–75
  mass, 29
  regional, 36–38, 40, 59, 63, 74–75
Flux measurement
  analytical bias, 71, 74
  confounded sources of variance, 74–75
  data analysis errors, 74–76
  diurnal variation, 112–115
  precipitation/irrigation effects, 70, 118
  random error vs. bias, 65, 75–76
  replicate observations, 63, 73–74
  sample handling bias, 71, 73–74
  sampling design errors, 74–76
  scaling issues, 63, 75, 103
  statistical tools, 65, 71–75
Footprint characteristics, 36, 55
Forests, 127, 129
  boreal, 174, 179
  temperate, 89, 174

tropical, 81, 89, 174
Fungal source of gaseous N oxides, 80

Gas(es)
  diurnal concentration variations, 36, 37, 39, 40
  diffusion in soil, 21, 65, 68, 83–84, 86–88
  foliar exchange, 63, 65
  greenhouse, 1–5, 19, 133, 157
    agricultural sources, 7–12, 45, 95–106
    mitigation strategies, 14–15
    projected changes, agriculture, 14–15
    nitrous oxide, 1–3, 5–11, 15, 45, 95–106
    carbon dioxide, 1–9, 45
    methane, 1, 3–10, 12–13, 15, 45
    CFC's, 1–3, 5–7
    ozone, 45
  horizontal soil transport, 69–70
  permeability, 154
  trace emissions (fluxes), 12–13, 95–106
General circulation models (GCM), 1
Global warming
  agriculture effect, 7–8
Global Warming Potential (GWP), 5–6
Gradient techniques, 22–26
Grassland(s), 39, 81, 89, 109
Grazing, 109, 116–118
Green manure, 183, 187–188, 191–192
Greenhouse effect, 1–3, 19, 157
Greenness index, 59
Groundwater, 99, 121, 122, 126–128, 129, 130
Gust probe, 49

Heterotrophic nitrification, 81
High-pass filter, 50
Hydrogenotrophs, 145
Hydroxylamine, 82
Hydroxylamine oxidoreductase, 82, 152

Industrial combustion, 97–98, 102
Infrared radiation, 59, 157
Inhibition, 82, 179, 180
Interfaces, 151
Interspecies H_2 transfer, 147
Isotope ratio, 193
  manually operated soil gas measurement, 194
  automated soil gas measurement, 196
  mass spectrometer, 193
Instrument response times, 50
Inversion, 34

K_m for methane, 150, 175

Labile carbon, 171
Lacunae, 154
Lag effects, 52, 50
Lagrangian
  dispersion, 30
  inverse, 30

Lagrangian (cont.)
  methods, 30–34
  time scale, 33
Lasers, 20, 24, 49

MMO, 148
  substrate versatility of, 148
Mass balance, 28
Mass spectrometers, 183–192, 193
Methane (CH$_4$), 1–10, 12–13, 157, 160,
    183–192
  affinity for, 150
  budget, 9
  consumption, 133, 135, 140–142, 148, 168,
    174
  emissions, 24, 161, 162, 163
  isotope measurement of, 197
  monooxygenase, 148
  production, 145
Methanogenesis, 158, 168, 169
  temperature, 147
Methanogens, 145
  pH tolerance of, 148
Methanotroph(s), 133, 148, 176, 181
  acid tolerant, 151
Methanotrophic nitrification, 151
Methylococcus capsulatus, 150
Methylotroph(s), 148
Microaerophilic methanotrophs, 151
Microclimate, 20, 21
Micrometeorological measurement/tech-
    nique/systems, 20, 22, 28, 34, 46, 54,
    63, 76
Mitigation, trace gas, 14–15
Mixing ratio, 23, 46, 54
Moisture, 179
Molybdate, 152
Momentum, 23, 34

Net primary productivity (NPP), 8
Nitrapyrin, 82, 135, 138–139, 152
Nitrate (NO$_3$), 80–90, 112–113, 117, 122, 124,
    126–128, 129–130, 137–139
Nitric oxide (NO), 3, 13, 63, 70, 79–80, 82–83,
    85–86, 88–90, 97, 102–106, 109–118
Nitrification, by methanotrophs, 151
Nitrite (NO$_2$), 115
Nitrogen dioxide (NO$_2$)
  oxidation, 81–82, 97, 102–106
  reduction by nitrifiers, 82
  toxicity, 82–83
Nitrogen oxides (NO$_x$), 19, 79–90, 97,
    102–106
  fungal source, 80
Nitrogen (N), 179
  availability, 109, 116–117
    to nitrifiers, 81, 83, 88–90
    to denitrifiers, 86–90
  cycle, 81, 85, 89, 98, 102, 116–118
  fertilizer, 98–102

  fixation, 9
  immobilization, 81
  mineralization, 81, 111, 116–117
  redistribution via the atmosphere, 79
  uptake by plants, 81, 86
Nitrogen gas (N$_2$)
  fixation by methanogens, 147
  fixation by methanotrophs, 150
  isotope measurement of, 194
Nitrous Oxide (N$_2$O), 1–3, 5–11, 15, 19–22,
    24, 26–28, 30, 34, 36, 39, 40, 47, 79–80,
    82–83, 85–90, 95–106, 121, 122,
    124–125, 126–130, 157, 193
  budget, 9
  degassing from storage vial septa, 198
  dinitrogen ratios, 86, 88, 138, 140
  emissions, 63, 67, 70, 133, 135–142
  isotope measurement of, 193
Nitric oxide (NO), 97, 102–106, 109–118, 194
Nitrification, 111–117, 140
  by methanotrophs, 151
  inhibitors, 82, 134–142
Nitrifiers, 133, 175, 178–180
Nitrobacter, 81–82
Nitrobacteriaceae, 81
Nitrogen dioxide, 80, 97, 102–106
Nitrosomonas, 81–82
Nitrosospira, 81
Nitrous acid disproportionation, 80
Nondenitrifying NO$_3$-reducing bacteria, 80,
    127
Non-methane hydrocarbons, 3, 13
N-fixing bacteria, 85

Oak trees, 109–110, 116–118
Oceans, 85
OH, 79, 97
Open chamber systems, 20, 21, 65
Open-path analyzers, 27
Organic matter, 158, 161
Oxygen (O$_2$), 169, 175, 180
  depletion, 159
  availability
    to nitrifiers, 81–83, 89–90
    to denitrifiers, 85–90
Ozone (O$_3$), 1, 13, 19, 59, 79, 95–97, 109, 118
  depletion, 157
  destruction, 95–96, 133

Passive sampler, 30
Peatland, 168
Permeability to gases, 154
pH, 171, 173
Photochemical reactions, 63, 71, 79
Planetary boundary layer, 34
Plant canopies, 30–34, 97
Plant exudates, 70
Plant gas transport, 70, 154, 168, 173
Population, increase, effects, 15
Potential temperature, 34

Primary production, 171, 173
Pseudomonas, 85, 88, 123

Q_{10}, 21
Quercus, 110

Radiation, 2
  net, 57
Radiative, balance, forcing, 2, 7
Redox, 124, 180
Reforestation, 14
Residue, crop, 89
Respiration, 178
Rhizosphere, 154, 173, 174
Rhodotorula, 151
Rice, 9, 15, 175, 180
  cultivation, 158
  flooded, 183, 188, 190–192
  methane emission, 24
  rhizosphere, 159
  straw, 161
Root exudates, 70, 86, 161, 162, 183, 191

Savannah, 81
Scale, 36
Scaling issues, 63, 75, 79, 88–90, 103–106
Sensor sensitivity, 68,76
Sensors
  fast-response, 49
  slow-response, 49
Sesbania aculeata, 184, 187, 191, 192
Sequential reduction sequence, 159
Slurries, 171
Soil
  carbon, 188, 189, 192
    loss, 8
    storage, 15
  compaction, 69–70
  drainage, 83, 89
  gas flux—calculation, 71–73
  microsite heterogeneity, 81, 89
  pH, 81, 83, 86, 158, 163
  redox potential, 124, 158, 159
  temperature, 21, 67, 112–115, 123, 160, 161
  texture, 89, 123, 124
  types, 158, 162
Sonic anemometers, 49
Sources and/or sinks, 19, 30–34
  processes, 63
Spatial scale, 74–75, 103–106
Spatial variability, 59, 63, 72, 74–75, 103–106
Standard deviation of vertical wind speed, 31, 33, 48
Stratified sampling approaches, 75, 89

Storage vials—for gas samples, 71, 198
Subsoil, 121, 122–125, 128, 129, 130
Sulfate (SO_4), 13–14, 152
Sulfate reducers, 159, 162
Sulfur dioxide (SO_2), 3, 19
Surface roughness, 23, 28, 29

Temperature, 186, 190–192
  arthopogenically-induced, 1–3, 6–7
  diurnal variations, 112–115
  global surface, 1–2
  methanogenesis, 147
  quotient, 147, 151
  stratification, 23, 28–30
Temporal variability, 63, 74–75
Terpenes, 19
Texture, 123, 124
Thermodynamics, 85
Thermophylic bacteria, 115
Tower-based, 56, 57, 63
Trace gas, 95–106
  absorption bands, 157
Tracer(s), 23
Transport, 169, 173, 174
Tunable diode laser, 49
Tundra, 173, 180
Turbulence, 30, 34
Turbulent diffusion, 22
Turbulent transfer, 46

Upland, 168, 174, 175, 180
Urea, 83, 162
  fertilization, 99, 134–142

Variability (of gas emissions), 22, 28, 40, 63, 103–106
Variance technique, 48

Water, 178
Water table, 174
Wetland(s), 70, 83, 85, 167–169, 173, 175, 180–181
  plants, 22
Wheat
  crop, 31–34, 37, 38
  dryland, 134–136, 140–141
  irrigated, 134–137, 140–141
Windspeed, 23–26, 28, 29, 31
  sonic anemometers, 49
Woodlands, 109

Yeasts, 80
  methanotrophic, 151